Applied and Numerical Harmonic Analysis

Gabor T. Herman

Geometry of
Digital Spaces

Birkhäuser
Boston • Basel • Berlin

Gabor T. Herman
MIPG, Department of Radiology
University of Pennsylvania
Philadelphia, PA 19104-6021

Library of Congress Cataloging-in-Publication Data
Herman, Gabor T .
    Geometry of digital spaces / Gabor T. Herman .
        p .   cm .
    Includes bibliographical references and index.
    ISBN 0-8176-3897-0. -- ISBN 3-7643-3897-0
        1 . Image Processing -- Digital techniques   2 . Geometry .
    3 . Computer graphics .   I. Title .
    TA1637 . H47
    621.36 ' 7--dc21   98-4736
                                                                        CIP

Printed on acid-free paper
© 1998 Birkhäuser Boston   *Birkhäuser*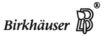

ISBN 0-8176-3897-0
ISBN 3-7643-3897-0

Printed in the U.S.A.
9 8 7 6 5 4 3 2 1

Contents

Preface

"La narración literaria es la evocación de las nostalgias."
("Literary narration is the evocation of nostalgia.")

G. G. Márquez, interview in *Puerta del Sol*, VII, 4, 1996.

A Personal Prehistory

In 1972 I started cooperating with members of the Biodynamics Research Unit at the Mayo Clinic in Rochester, Minnesota, which was under the direction of Earl H. Wood. At that time, their ambitious (and eventually realized) dream was to build the Dynamic Spatial Reconstructor (DSR), a device capable of collecting data regarding the attenuation of X-rays through the human body fast enough for stop-action imaging the full extent of the beating heart inside the thorax. Such a device can be applied to study the dynamic processes of cardiopulmonary physiology, in a manner similar to the application of an ordinary CT (computerized tomography) scanner to observing stationary anatomy.

The standard method of displaying the information produced by a CT scanner consists of showing two-dimensional images, corresponding to maps of the X-ray attenuation coefficient in slices through the body. (Since different tissue types attenuate X-rays differently, such maps provide a good visualization of what is in the body in those slices; bone — which attenuates X-rays a lot — appears white, air appears black, tumors typically appear less dark than the surrounding healthy tissue, etc.) However, it seemed to me that this display mode would not be appropriate for the DSR. If that device allowed us to display the distribution of the X-ray attenuation coefficient in 30 slices through the heart averaged over a sixtieth of a second, then two seconds worth of data would result in 3,600 two-dimensional images. Who would be willing to study them all and who would be capable of mentally integrating the four-dimensional (i.e., time-varying, three-dimensional) information in them?

For these reasons, I proposed to H. K. Liu (at the time a graduate student in the Department of Computer Science, State University of New York at Buffalo, where I was on the faculty) that he work with me for his doctoral dissertation on the problem of creating computer graphic displays of the time-varying three-dimensional appearance of the beating heart from the four-dimensional data produced by the DSR. This he successfully completed and by 1976 a movie was created showing the changing appearance of a moving cardiac surface from user-specified directions. Earl Wood referred to what we were doing as "noninvasive vivisection."

An essential part of the total process in producing such movies was the extraction of the surfaces to be displayed from the four-dimensional data set. H. K. Liu solved this problem by creating the whole time-varying surface by locally following maximal changes in the four-dimensional data set (on the basis that the changes in values across the surface of an organ are likely to be larger than just inside it or just outside it) and having a backtracking mechanism to explore alternative directions of search if evidence indicated that the program got lost. This worked well, but took a very long time to produce its output.

A computationally much faster procedure was proposed to us by Ehud Artzy (who joined our research group having obtained his Ph.D. in our department by solving a problem of graph theory). The basic idea was that if we can first segment the data set so that in place of the originally estimated X-ray attenuation coefficients we have either a 1 or a 0, with 1 indicating that the corresponding region of space is occupied by the desired type of tissue, then the problem of producing the desired surfaces reduces to that of tracking the boundary of a connected component of 1s. Artzy transformed this problem into one of efficiently traversing all nodes in a component of a graph (I discuss these ideas in detail in the body of this book) and proposed an algorithm to do that.

This algorithm was implemented and seemed to work on all examples on which it was tested, but we found it surprisingly difficult to prove that it would work correctly for an arbitrary three-dimensional arrangement of 0s and 1s. Artzy's office was full of piles of sugar cubes, representing various possible arrangements of 1s in space. In trying to get an insight into a valid proof, all sorts of physical pictures of the behavior of the algorithm had been created, including a description of it as a "green liquid flowing on the surface" (by Gideon Yuval). My own image for explaining the procedure was invented for a talk I gave at a conference organized by Karl-Heinz Höhne in Hamburg; it uses the cloning flies which form the topic of the first chapter of the book. The algorithm was published in the archival literature in 1981, with a less than satisfactory outline of its proof of correctness.

It appeared to us that the field of mathematics that might supply us with appropriate theoretical underpinnings would be topology. However, the search for known topological approaches and results which would easily yield our desired proof of correctness of Artzy's algorithm turned out to be disappointing. The essential problem was that results of classical topology let us down at exactly those places where we had difficulties in our initial attempts at a proof. These were mainly caused by the discrete nature of our data and our output. (One aspect of this is that the surfaces produced by the algorithm may touch themselves — mathematically speaking, they are not necessarily 2-manifolds — even though they have perfectly well defined insides and outsides in the appropriate discrete interpretation.) Eventually (in 1983) a mathematically rigorous proof of correctness was published (with the help of D. Webster). This proof, although it is basically topological, is ad hoc and complicated; it certainly lacks mathematical elegance. This is still caused by the inappropriateness of classical continuous topology for the study of the discrete objects in digital spaces; much of the motivation for this book comes from the desire to have a body of mathematical knowledge which is directly relevant to such applications.

Aim and Scope of This Book

The desire to provide a solid mathematical theory which is appropriate for discussing geometrical concepts as they occur in the digital environment has resulted over the last twenty years in a sizable body of published research. This book provides an introduction and overview of this literature. Certainly, the proof of correctness and the analysis of the properties of the output of Artzy's algorithm are trivial consequences of the general theory presented in the body of our text, but there is much more. The intent is to provide a self-contained and mathematically pleasing body of knowledge, which can immediately be used in the discussion of properties of practical algorithms in digital spaces.

The *Geometry of Digital Spaces* is not as yet a well-established subject and so there is no standard set of topics that an author is expected to cover in a book devoted to it. Naturally, I concentrate on those specific parts of the theory which I found most useful in the applications that I happened to have pursued. However, what I consider most important in the book is not the exact list of results, but rather the illustration that it is possible to develop a geometry of digital spaces which achieves the dual aim of being both mathematically rigorous and immediately applicable to practical problems. It is my experience that the material presented in the book can be covered in a one-semester course for advanced undergraduate students.

For whom is this book written? There are two groups of people who may find its content both educational and fun: (i) practitioners of image processing and computer graphics and (ii) mathematicians. The first group will find a wealth of useful concepts and results which will allow them to analyze their computer algorithms for processing multidimensional data in better ways than they may have thought possible. They may, if they wish, take my results on faith and skip the proofs. They are mainly there for the mathematical audience; this book is offered to them as a new geometry, one that is appropriate to the digital world forced upon us by our computers. Because of this diversity of the proposed audience, no knowledge is taken for granted and all concepts are carefully introduced, defined, and illustrated. I have much enjoyed writing this book. I hope that I have not failed in conveying my delight in its subject matter.

I wish to make some comments regarding the exercises. For any reader on top of the subject, they should not be difficult to do; hence they are there so that the reader can check whether or not the material of the previous chapter has been mastered. Also, they are not irrelevant; on the contrary, it will often be the case that the conclusion of an exercise in an earlier chapter is assumed to be proven in a later chapter. The reader should also note that in the Appendix there is a list of the symbols used in the book together with references to the places where the symbols are defined.

Acknowledgments

Over the years, my work on the geometry of digital spaces greatly benefited from three individuals. With two of them, Jayaram Udupa and Yung Kong, I had many discussions on various aspects of the field; many of the specific ideas which found their way into the book have originated from these discussions. On the other hand, I have never had the privilege of working closely together with Azriel Rosenfeld. Nevertheless, he is clearly the intellectual grandfather of some of my approaches.

I have been fortunate in having had support for my research throughout the period described in this preface, mainly from the National Institutes of Health and from the National

Science Foundation. Most relevantly, I have been given support to develop an advanced undergraduate course on the Geometry of Digital Spaces from a National Science Foundation grant on Mathematics and its Applications Across the Curriculum (DUE9552464, Principal Investigator Dennis DeTurck). This grant supported several student assistants (Edgar Garduño, Kelly O'Leary, Daniel Shub, and Eric Tsai); interactions with them regarding preliminary versions of this book have greatly improved the presentation of the material. Samuel Matej was a helpful faculty collaborator on the grant. Additional help was provided by Bruno Motta De Carvalho, T. N. Jones, Vera Peshchansky, and Avi Vardi.

This has been my first attempt to prepare a "camera-ready" version of a book on a computer. In this I have been greatly helped by Wayne Yuhasz of Birkhäuser.

Finally, I wish to thank my wife, Marilyn Kirsch, who not only put up with the many weekends and late evenings that I spent working on this book, but supported the idea with great enthusiasm and even contributed by producing the original drawing of the Fat Fly. Every scientist needs an artist in the family!

Gabor T. Herman
Philadelphia, March 13, 1998

1
Cloning Flies on Sugar Cubes

"A fly can't bird, but a bird can fly."

A. A. Milne, *Winnie-the-Pooh*, Chapter VI.

1.1. What Is Our Game?

Many imaging devices will produce estimated values of a physical quantity at certain points (referred to as *grid points*) in three-dimensional space. For example, a computerized tomography (CT) scanner estimates the X-ray attenuation coefficient inside a human body at points of a three-dimensional rectangular grid. When displaying the results of such an estimation, we usually use a sequence of two-dimensional images, such as those shown in Figure 1.1.1. In these images the bone in the spine and in the ribs appears light (the more concentrated the bone is, the lighter it appears), softer tissues appear gray, and the air outside the body and in the lungs appears dark.

Such displays are obtained in the following fashion. Let us assume that the X-ray attenuation coefficients have been estimated at grid points with integral coordinates (c_1, c_2, c_3), $1 \leq c_1 \leq I$, $1 \leq c_2 \leq J$, $1 \leq c_3 \leq K$. (Here, K is the number of slices, such as shown in Figure 1.1.1, I is the number of rows and J is the number of columns in a slice.) We fix c_3 to be k and consider the set of points $S_k = \{(c_1, c_2, k) \mid 1 \leq c_1 \leq I, 1 \leq c_2 \leq J\}$.

Now we turn aside for a moment to introduce some concepts and notations that will be used throughout our text. We use R to denote the set of real numbers and R^N to denote the set of N-dimensional (row) vectors all of whose components are real numbers. If $v \in R^N$, v_n denotes the nth component of v, for $1 \leq n \leq N$. The *inner product* of two such vectors u and v is defined as

$$u \cdot v = \sum_{n=1}^{N} u_n v_n , \qquad (1.1.1)$$

and the *norm* of a vector v is defined as $\|v\| = \sqrt{v \cdot v}$. As usual, we interpret the components of an element of R^N as the coordinates of a point in a rectangular coordinate system in an N-dimensional euclidean space (i.e., in a plane if $N = 2$ or in the ordinary three-dimensional space if $N = 3$). In fact, we will (in accordance with common practice) use the algebraic notion of a *vector* in R^N and the geometric notion of a *point* in an N-dimensional euclidean

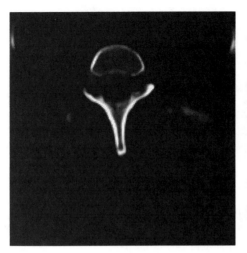 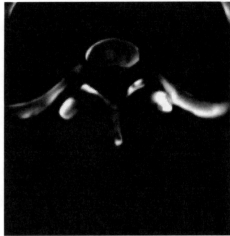

Figure 1.1.1. Two slices of a CT scan.

space quite interchangeably, to the extent of actually referring to R^N as **the** N-dimensional euclidean space. Under this identification, the norm of a point is its distance from the *origin* (i.e., from $(0, \cdots, 0)$) of the coordinate system.

We use Z to denote the set of all integers. For any positive real number δ and for any positive integer N, we define

$$\delta Z^N = \{ (\delta c_1, \cdots, \delta c_N) \mid c_n \in Z, \text{ for } 1 \le n \le N \} . \qquad (1.1.2)$$

Specifically, we use Z^N to abbreviate $1Z^N$. (The set-theoretical notation we use in this and in all following definitions is standard. In particular we use $\in$ for "is an element of" and $\notin$ for "is not an element of .")

Now let G be any set of points in R^N. (In this context we refer to G as a *grid* in R^N.) The *Voronoi neighborhood* in G of any element g of G is defined as

$$N_G(g) = \{ v \in R^N \mid \text{for all } h \in G, \ \|v - g\| \le \|v - h\| \} . \qquad (1.1.3)$$

In words, the Voronoi neighborhood of g consists of all those vectors which are not nearer to any other point of G than they are to g. For example, for $\delta > 0$, the Voronoi neighborhood of $(\delta c_1, \delta c_2)$ in δZ^2 is $\{ (v_1, v_2) \in R^2 \mid \max(|v_1 - \delta c_1|, |v_2 - \delta c_2|) \le \frac{\delta}{2} \}$. This is a closed square, i.e., it contains its edges (an alternative common terminology for "edge" is "side"), with side-length δ centered at the point $(\delta c_1, \delta c_2)$. When perceived as a set of points in R^2, δZ^2 is referred to as a *square grid*. The square grid Z^2 and one of the associated Voronoi neighborhoods are illustrated in Figure 1.1.2. The Voronoi neighborhoods in a grid in R^2 are referred to as *pixels* (short for picture elements).

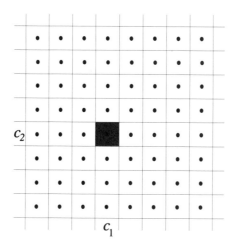

Figure 1.1.2. The square grid Z^2 and one of its Voronoi neighborhoods (a pixel).

There are certain restrictions that one may usefully apply to the concept of a grid without eliminating any grids which are of interest. For example, for any positive real numbers D and d, we refer to a grid G in R^N as a (D, d) *grid* if it satisfies the following two properties.

    (i) For all $v \in R^N$, there exists a $g \in G$ for which $\|v - g\| \leq D$ (i.e., the grid points are not "too sparse" anywhere).

    (ii) For all $g \in G$, there is no $h \in G$ for which $0 < \|h - g\| < 2d$ (i.e., the grid points are not "too dense" anywhere).

Following [40], we refer to a grid G in R^N which is a (D, d) grid, for some positive D and d, as a *Delone grid*. We leave it to the reader to prove that, for any positive integer N and for any positive real number δ, δZ^N is a Delone grid.

Now we return to the way the two-dimensional images of Figure 1.1.1 are produced. The region over which the images are defined is the union of the pixels associated with those

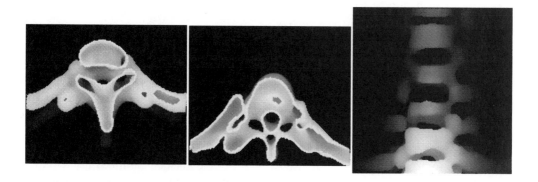

Figure 1.1.3. Computer graphic displays of a detected boundary of the spine and the attached ribs, based on the data set which contains the slices shown in Figure 1.1.1.

$(c_1, c_2) \in Z^2$ for which $1 \leq c_1 \leq I$ and $1 \leq c_2 \leq J$. The gray value assigned to the pixel associated with one of these (c_1, c_2)s is proportional to the estimated X-ray attenuation coefficient at (c_1, c_2, k). (For display purposes, there may be some cropping of the gray values so that they lie in the range [0, 255].) This is the image of the slice containing the points of S_k.

The overall aim of the type of work for which we are developing the mathematical theory in this book is the production of computer graphic displays of selected objects based on the estimated values of the physical quantity at grid points in three- (and sometimes higher) dimensional space. Examples of such displays of a spine and of some of the attached ribs are shown in Figure 1.1.3. These have been obtained from the three-dimensional data set which contains the slices of Figure 1.1.1.

1.2. A Methodology for Extracting Object Boundaries

In this section we discuss a particular methodology for extracting boundaries of objects based on values assigned to points of the grid Z^3 in R^3. For $\delta > 0$, δZ^3 is referred to as a *cubic grid*. The Voronoi neighborhoods associated with a grid in R^3 are referred to as *voxels* (short for volume elements). The voxel associated with the element (c_1, c_2, c_3) of Z^3 is the closed unit cube $\{ v \in R^3 \mid \max(|v_1 - c_1|, |v_2 - c_2|, |v_3 - c_3|) \leq \frac{1}{2} \}$. The tessellation of R^3 into voxels of a cubic grid is sometimes referred to as a *cuberille*, by analogy with the tessellation of the plane into pixels of a square grid which is like the quadrille ruling on paper (see Figure 1.1.2). In Figure 1.2.1, we illustrate a grid point g of Z^3 together with all other grid points which lie on the $2 \times 2 \times 2$ cube whose center is g (some or all of these may be considered "adjacent" to g, a concept which will be essential later in this book), as well as the Voronoi neighborhood of g.

Now suppose that we have some methodology to determine (based on the values assigned to the grid points) which grid points belong to what kind of material. For example, in CT bone attenuates X-rays more than any other type of tissue in the human body and so we may say that if the value assigned to a grid point (which is an estimate of the X-ray attenuation coefficient) is above a certain value, then that grid point is in bone. So, as a first and rough approximation (which will serve us for the discussion in this chapter), the part of space occupied by bone may be considered to be the union of the (cube-shaped) voxels associated with those pixels for which the estimated X-ray attenuation coefficient is above the threshold for bone. This collection of voxels may not necessarily form a connected subset of three-dimensional space: some pieces of bone may be disconnected from the rest and some (usually isolated) voxels may be included because noise in the estimation process caused us to identify a grid point as being in bone, when in fact it is not. Let us say that the "object" whose boundary we wish to display is a *component* of the set of points occupied by the "bone" voxels (i.e., it consists of a connected union of "bone" voxels, and it is not a subset of a larger connected union of "bone" voxels).

Even this specification is not precise enough to describe the intuitive notion of "a boundary" of an object. This is because an object of the kind we described in the previous paragraph may have holes inside it (just look at Figure 1.1.1; the inside of bones contains lots of less X-ray attenuating tissue and may well be identified as a result of thresholding as "not bone") and so it may have multiple boundaries (one exterior one and possibly many interior ones). Let us say that our task is to identify exactly one of these boundaries.

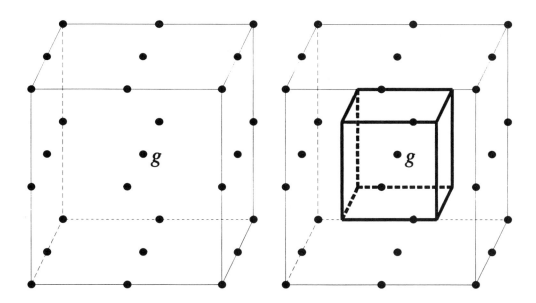

Figure 1.2.1. On the left we show a point g of the cubic grid Z^3 together with the other 26 grid points which lie on the $2 \times 2 \times 2$ cube whose center is g. On the right we show (using heavy lines) the Voronoi neighborhood of g (a voxel).

How do we specify which one? One way is to point at a boundary face, that is at a face which separates a bone voxel from a not bone voxel, and say that we wish to display that boundary of the object which contains that boundary face. (At this point, it is not even clear that this specification is legitimate. Is the "boundary containing a boundary face" a well defined concept? In what follows we will show that it is; we will set up a mathematical environment in which for any boundary face there will be one and only one boundary containing that face. For now we need to keep our discussion at an intuitive level and assume without worrying much about it that the terms we use have physical meanings.)

So the picture that we have is the following. We consider the cuberille, the tessellation of three-dimensional space into (cubic) voxels. A finite number of these voxels are occupied by sugar cubes of just the right size. We point at an uncovered face of one of these sugar cubes (i.e., a face such that the voxel on the other side of it is not occupied by a sugar cube) and ask that there be delivered to us the boundary surface which contains that face. The problem is to design a simple computing device (or program) which is guaranteed to do this for all possible arrangements of the sugar cubes. Before attacking the three-dimensional problem, we consider its analog in the simpler case of two dimensions.

1.3. Flies in Flatland

Just for now, we follow Mr. Abbott and refer to the two-dimensional plane as Flatland [1] and to the computing device which we hope will deliver us the required boundary surfaces

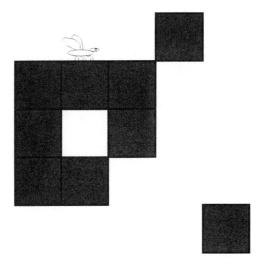

Figure 1.3.1. Sugar cubes in Flatland with a Flat Fly on a flat face (i.e., on an uncovered edge of an occupied pixel).

as a Flat Fly. As an example, consider a configuration of two-dimensional sugar cubes in Flatland shown in Figure 1.3.1 with a Flat Fly on one of the flat faces (i.e., on an edge) of one of these flat sugar cubes.

Algorithm for Flat Flies
(1) Dirty the face on which you are standing.
(2) Crawl onto the face which meets the one on which you are standing at the vertex in front of you.
(3) See if it is dirty.
    a.   If it is, fly away.
    b.   If it is not, start again at Instruction (1).

This algorithm actually does something that can be made mathematically precise. To state the essence of this book in less than ten words, it is about "making precise what algorithms like the one above do." To achieve that precision, we need to introduce some mathematical terminology. For now we simply demonstrate, in Figure 1.3.2, the behavior of the Flat Fly when put into the situation depicted in Figure 1.3.1.

1.4. Components Determined by Binary Relations

Now we introduce some very basic mathematical concepts and terminology; we will need these for the precise specification of the behavior of algorithms, such as the one for Flat Flies. In this section we illustrate all these concepts on the square grid Z^2, but further along in the book we will be applying the same ideas in much more general situations.

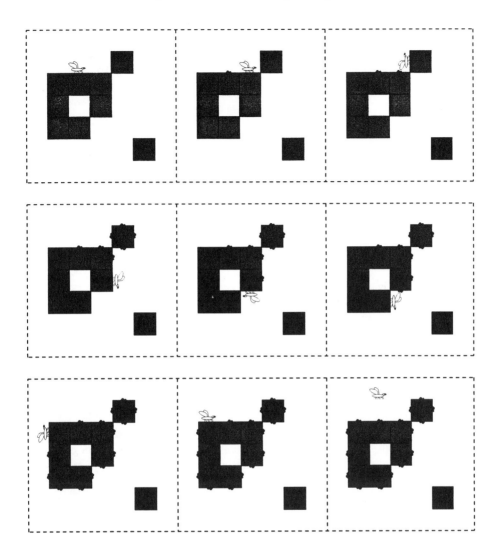

Figure 1.3.2. Illustration of boundary tracking in two dimensions by the Algorithm for Flat Flies: the first row is the beginning, the second row is the middle, and the third row is the end of the execution of the algorithm.

Let M be any set and ρ be a *binary relation* on M (i.e., ρ is a subset of M^2, the set of all ordered pairs of elements of M). If $(c,d) \in \rho$, then we say that c is *ρ-adjacent to d* and that d is *ρ-adjacent from c* and, in case ρ is a *symmetric relation* (meaning that $(c,d) \in \rho$ if, and only if, $(d,c) \in \rho$), that c and d are *ρ-adjacent*. We will often use the word "adjacency" to refer to a symmetric binary relation.

When $M = Z^N$, we will be repeatedly dealing with two binary relations, α_N and ω_N, defined as follows. (Both here and later, we use the symbol $\Leftrightarrow$ to abbreviate the phrase "if, and only if.")

$$(c,d) \in \alpha_N \Leftrightarrow (c \neq d \text{ and, for } 1 \leq n \leq N, \ |c_n - d_n| \leq 1) . \tag{1.4.1}$$

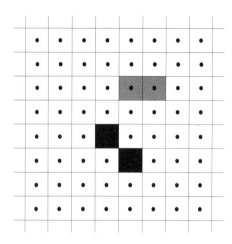

Figure 1.4.1. The grid points of the darker shaded pixels are α_2-adjacent but not ω_2-adjacent. The grid points of the lighter shaded pixels are both α_2-adjacent and ω_2-adjacent.

$$(c, d) \in \omega_N \Leftrightarrow \sum_{n=1}^{N} |c_n - d_n| = 1 . \tag{1.4.2}$$

Note that, for any positive integer N, $\omega_N \subset \alpha_N$. (We use $A \subset B$ to denote the phrase "the set A is a subset of the set B"; when we wish to express the idea that A is a *proper subset* of B — meaning that B contains an element not in A — we will explicitly say so.)

When $N = 2$ (a square grid), $(c, d) \in \alpha_2$ means that the associated pixels (Voronoi neighborhoods) have a nonempty intersection but are not identical, and $(c, d) \in \omega_2$ means that the associated pixels have exactly one edge in common. These ideas are illustrated in Figure 1.4.1. These are very often used notions in the literature and go under a number of commonly used names. For obvious reasons, ω_2 is referred to both as the edge-adjacency and as the 4-adjacency, and α_2 is referred to both as the edge-or-vertex-adjacency and as the 8-adjacency.

Let A be a subset of M. For any c and d in A, the sequence $\langle c^{(0)}, \cdots, c^{(K)} \rangle$ of elements of A is said to be a ρ-*path in A connecting c to d*, if $c^{(0)} = c$, $c^{(K)} = d$ and, for $1 \leq k \leq K$, $c^{(k-1)}$ is ρ-adjacent to $c^{(k)}$. We call K the *length* of this path. Note that the length is measured not by the number of elements in the path but rather by the number of steps needed to get from its beginning to its end. In particular, there are ρ-paths of length zero (such as $\langle c \rangle$); we refer to them as *trivial paths*. If there is a ρ-path in A connecting c to d, then we say that c is ρ-*connected in A to d*. (We note that ρ-connectedness in A is a reflexive relation, since every element of A is ρ-connected in A to itself by a trivial path.)

Let $M = Z^2$, and let A consist of all those grid points whose pixels are painted gray in Figure 1.4.2. We see that d is both ω_2-connected and α_2-connected in A to c, but the length of the shortest ω_2-path is five and the length of the shortest α_2-path is two. Also, e is α_2-connected, but not ω_2-connected, in A to c (or to d). On the other hand, f is neither ω_2-connected nor α_2-connected in A to c (or to d or e).

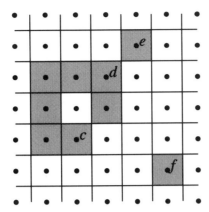

Figure 1.4.2. Illustration of ω_2-connectedness and α_2-connectedness (see text).

A nonempty subset A of M is said to be a *ρ-connected subset* if, for any c and d in A, c is ρ-connected in A to d. We see that the A which consists of all those grid points whose pixels are painted gray in Figure 1.4.2 is neither ω_2-connected nor α_2-connected. If we remove f from A, then what is left is α_2-connected but is not ω_2-connected. If we also remove e, then the remaining subset of Z^2 is both ω_2-connected and α_2-connected.

We have already seen that ρ-connectedness of elements in A is a reflexive relation. It is also a transitive relation, since if there is a ρ-path in A connecting c to d and a ρ-path in A connecting d to e, then they can be combined into a ρ-path in A connecting c to e. If it is also the case that ρ-connectedness in A is a symmetric relation on A (and hence, by definition, an *equivalence relation* on A), then it partitions A into *ρ-components* (i.e., into nonempty ρ-connected subsets which are not proper subsets of any other ρ-connected subset of A). If ρ happens to be symmetric, then ρ-connectedness in A is guaranteed to be a symmetric (and hence an equivalence) relation on A.

Referring to Figure 1.4.3, we see that there are three ω_2-components of the set of grid points whose pixels are painted gray: one has seven elements (all marked by an *) and the other two have one element each (marked by e and f, respectively). On the other hand, there are only two α_2-components, the one consisting of f alone and the other containing the rest of the shaded grid points.

We denote the *complement* $M - A$ of A in M by $\overline{A}$. In the case of Figure 1.4.3, $\overline{A}$ (the set of grid points whose pixels are not painted gray) is an α_2-connected subset of Z^2 (and hence it has only one α_2-component, namely, itself). On the other hand, it is not an ω_2-connected subset of Z^2, since the grid point marked by an # is not ω_2-connected in $\overline{A}$ to any other grid point in $\overline{A}$ and hence forms an ω_2-component of $\overline{A}$ by itself. The remaining elements of $\overline{A}$ form another (the only other) ω_2-component of $\overline{A}$.

We complete this section by stating a terminological convention which we will be using repeatedly later on. In the special case when $A = M$, we use the phrases *ρ-path* and *ρ-connected* instead of ρ-path in A and ρ-connected in A. Thus, in Figure 1.4.2, e is ω_2-connected to c, even though e is not ω_2-connected in A to c. (It is indeed the case that every grid point in Z^2 is both ω_2-connected and α_2-connected to every other grid point in Z^2.)

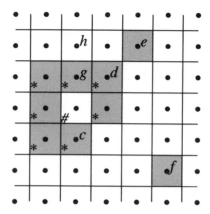

Figure 1.4.3. Components in sets of pixels (see text).

1.5. So, What Does a Flat Fly Do?

Now we have all the necessary terminology at hand to specify the outcome of the Algorithm for Flat Flies. First let us look at the specific case identified in Figures 1.3.1 and 1.4.3.

As we can see, the Flat Fly is initially put on the face which separates the grid point g (painted gray) from the grid point h (not painted gray). After a while it flies away (see Figure 1.3.2), leaving behind a bunch of faces that have been dirtied. This set of faces can be described as consisting of exactly those faces which are between a grid point painted gray which is α_2-connected in the set of gray grid points to g and a grid point not painted gray which is ω_2-connected in the set of not gray grid points to h.

In fact this is representative of the general behavior of the Flat Fly. We can say the following. Suppose that in the square grid Z^2 (see Figure 1.1.2) the pixels of a finite number of grid points are filled with flat sugar cubes (are painted gray, if you prefer to say it that way) and that a Flat Fly is put on top of one of these sugar cubes into a pixel which is not occupied by a sugar cube (see Figure 1.3.1). Let g be the name of the grid point of the sugar cube on top of which the Flat Fly is placed into the pixel of a grid point named h (see Figure 1.4.3).

Claim 1.5.1. *After a finite number of loops through Instructions (1), (2), and (3) of the Algorithm for Flat Flies, the Flat Fly will fly away (because Condition (3)a is applicable). A face will be dirty at this time if, and only if, it has the property that it is a face between a pixel associated with an element of the α_2-component of the set of grid points with sugar cubes that contains g and a pixel associated with an element of the ω_2-component of the set of grid points without sugar cubes that contains h.*

We do not give a proof of this claim in this chapter; such a proof will be a consequence of the general results that we will be deriving in the chapters which follow. We will return to proving the claim at the appropriate point later in the book. However, before going on to other things we state another claim which shows that the set of faces dirtied by the Flat Fly has some desirable properties. First we need to introduce some additional notation and terminology.

First we observe that the face onto which the fly is initially placed can be identified by the pair (g, h) of grid points on either side of it. Note that this is an ordered pair, interpreted as "the fly is put into h, with its feet touching g." If we were to draw the Flat Fly on (h, g), then its feet would be touching the same edge, but it would be hanging upside down. Similarly, every ordered pair of ω_2-adjacent grid points can be thought of as such an edge (two-dimensional face) with an orientation across it from the first grid point in the pair toward the second and, conversely, all such oriented edges can be described by a unique element of ω_2. Thus we make the fundamental observation (analogs of which will be used all through this book) that ω_2 has a dual purpose: it is an adjacency on Z^2 and its elements can be used as unique identifiers of elements of the set of all oriented faces between pixels of this square grid. Immediately we note that, for an arbitrary positive integer N, ω_N is an adjacency on the grid Z^N and its elements can be used as unique identifiers of elements of the set of all oriented (hyper-)faces between points of the grid. Now we introduce two concepts for the grid Z^N. These will be generalized to arbitrary digital spaces later on.

Let O and Q be two subsets of Z^N. We define the *boundary between O and Q* as

$$\partial(O, Q) \ = \ \{\, (c, d) \,|\, c \in O, \, d \in Q, \, (c, d) \in \omega_N \,\}. \tag{1.5.1}$$

(Do not confuse the boundary symbol ∂ with the Greek letter δ which is used for other purposes in this book.)

Let $\langle c^{(0)}, \cdots, c^{(K)} \rangle$ be an ω_N-path and S be any subset of ω_N. We say that $\langle c^{(0)}, \cdots, c^{(K)} \rangle$ *crosses* S, if there is a k, $1 \le k \le K$, such that either $(c^{(k-1)}, c^{(k)}) \in S$ or $(c^{(k)}, c^{(k-1)}) \in S$.

Claim 1.5.2. *Let A be any nonempty proper subset of Z^2. Let O be an α_2-component of A and Q be an ω_2-component of $\overline{A}$, such that $\partial(O, Q)$ is not empty. Then there exist two uniquely defined subsets I and E of Z^2, which have the following properties.*

  (i) *$O \subset I$ and $Q \subset E$.*
  (ii) *$\partial(O, Q) = \partial(I, E)$.*
  (iii) *$I \cup E = Z^2$ and $I \cap E = \emptyset$. ($\emptyset$ denotes the empty set.)*
  (iv) *I is an α_2-connected subset of Z^2 and E is an ω_2-connected subset of Z^2.*
  (v) *Every ω_2-path connecting an element of I to an element of E crosses $\partial(O, Q)$.*

The proof of this claim is also delayed till later; in fact it will be proven in a much more general context. Here we discuss its relevance with respect to the Algorithm for Flat Flies.

Let A consist of the nonempty finite set of grid points which have been painted gray according to the supposition preceding Claim 1.5.1, and let the Flat Fly be placed on (g, h), where $g \in A$ and $h \in \overline{A}$. Let O be the α_2-component of A containing g and Q be the ω_2-component of $\overline{A}$ containing h. Then, in our newly introduced terminology, Claim 1.5.1 says nothing more or less than that the set of faces dirtied by the Flat Fly is exactly $\partial(O, Q)$.

Let us check on this for our previous example. Given the situation shown in Figure 1.4.3, we see that O is the set of all grid points marked by an * together with the grid point marked by e and Q is the set of grid points not painted gray except for the one marked by #. We see that the output (i.e., the dirtied faces) of the Algorithm for Flat Flies (see Figure 1.3.2) is indeed $\partial(O, Q)$.

Now let us see what Claim 1.5.2 has to say about this boundary. It says that there are two sets of grid points I and E, of which we will think intuitively as the interior and the exterior, which have certain properties. Now we specify what these sets are for the special case depicted in Figure 1.4.3; it is easy to check that the specified sets indeed have all the required properties. The interior for the case of Figure 1.4.3 consists of all grid points marked by an * together with the two grid points marked by e and by #, respectively. The exterior consists of all the grid points not in the interior.

Part of what is stated above is reminiscent of the famous Jordan Curve Theorem [31]. Since much of what we do is inspired by this theorem, now we give an intuitive description of it. (We do not give a formal statement or proof, first, because we do not make any formal use of the theorem in our theory and, second, since a rigorous proof of the theorem from first principles is quite complicated.)

The Jordan Curve Theorem is about properties of simple closed curves in the plane R^2. Intuitively, a *simple closed curve* is one that can be drawn by putting a pencil on a sheet of paper and continuously drawing with it without ever again visiting a point that was previously drawn (this is *simplicity*) right until we get back to the starting point (this is *closedness*). An example of such a curve is shown on the left side of Figure 1.5.1. The Jordan Curve Theorem says that the set of all points in the plane which are not on a simple closed curve can be partitioned into two subsets, one may be called the interior and the other the exterior (see the right side of Figure 1.5.1). Both of these are connected sets in the sense that we can get from any point of the interior to any other point by drawing continuously a curve between them which never leaves the interior (and similarly for the exterior), but one cannot draw continuously a curve from a point in the interior to a point in the exterior which does not contain at least one point of the simple closed curve, which is in fact the boundary between the interior and the exterior. (Here we use the word "boundary" in its traditional continuous sense: the boundary between the interior and the exterior is the set of those points for which every open disc centered at the point also contains points both from the interior and from the exterior.)

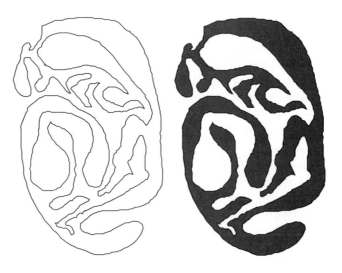

Figure 1.5.1. Example of a simple closed curve (on the left) with its interior shaded gray (on the right).

The boundary that is the output of the Algorithm for Flat Flies shares important properties with simple closed curves in R^2: its interior and exterior partition the whole space (iii), both are connected in some sense (iv), but one cannot get from the interior to the exterior without crossing the boundary (v) which, in fact, is the boundary between the interior and the exterior (ii). (Mathematically speaking there is also a difference: a simple closed curve is a subset of the plane, and the Jordan Curve Theorem says that the complement of the curve, rather than the whole plane, has precisely two components. Since our boundary is a subset of ω_2, rather than of Z^2, it is possible to demand that the interior and exterior partition the whole of Z^2, if anything, a more attractive-sounding aim than that of the classical theorem.)

It is important to recognize that, in spite of this formal similarity to the Jordan Curve Theorem, Claim 1.5.2 is not a result about the continuous plane R^2 but a thoroughly digital theorem. Notions of "connectedness," "path," and "crosses" have all been defined for the square grid Z^2 and although they are analogous to corresponding notions in R^2, sometimes the analogy breaks down. It is the very difficulty of applying the continuous notions directly to the digital environment to prove claims (such as the two above) regarding algorithms operating in discrete domains which motivated us to develop a geometry directly for digital spaces.

For example, it is tempting to reinterpret the boundary output by the Algorithm for Flat Flies as the point set in R^2 which is the union of all the points in all the dirtied faces. However, in spite of the discrete Jordan curve properties of this boundary provided to us by Claim 1.5.2, the Jordan Curve Theorem is not applicable to the corresponding point set in R^2, since it does not form a simple closed curve. (This is because it "touches itself" at the vertex shared by the pixels d and e in Figure 1.4.3, and so it is not a simple curve.)

Having said this, we must admit that the argument is only half convincing. It is all very well that we can give a self-contained theory in digital spaces with nice theorems which resemble those of famous results of continuous mathematics. Nevertheless the reinterpretation in the underlying continuous space is unavoidably needed for some applications. For example, no physician would think of the boundaries shown in Figure 1.1.3 as collections of ordered pairs of voxels. Rather they would be interpreted as (digital approximations of) continuous biological surfaces. For this reason, it is not really desirable to have boundaries which touch themselves in the fashion of the boundary consisting of the dirtied faces at the end of Figure 1.3.2.

While we are pondering this, we may as well ask the following question. Is it (biologically, physically, whateverly) reasonable that the pixel e is considered connected to the set of pixels marked by an * in Figure 1.4.3? This came about quite naturally in the way the Flat Fly crawled about on the sugar cubes in Figure 1.3.2, but, looking at the result, the connection is rather suspect. Would it not be better to use only ω_2 to define connectivity and avoid such flimsy looking connections?

Once we raise this objection on physical grounds, we also see that the use of α_2 is not particularly elegant even from a mathematical point of view. Why do we use different types of components for grid points with sugar cubes and for those without in Claim 1.5.1? Similarly, why do we use different types of components for A and for $\overline{A}$ in Claim 1.5.2? Since in (v) of Claim 1.5.2 only ω_2 is used, would it not be much nicer to replace α_2 by ω_2 everywhere in Claim 1.5.2?

Of course it would be nicer, but the result would not be true. To show this, we need to look at an example other than the one of Figure 1.4.3. There, the boundary between the ω_2-component of A containing g (the set of all grid points marked by an *) and the ω_2-component of $\overline{A}$ containing h has an interior (the set of all grid points marked by an *

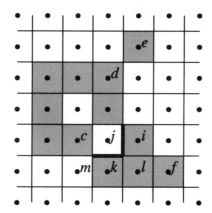

Figure 1.5.2. Illustration that ω_2-connectedness cannot be used for both the sugar cubes and their background.

together with the one marked by #) and an exterior which satisfy all the properties listed in Claim 1.5.2. However, look at Figure 1.5.2.

To show that Claim 1.5.2 becomes false if everywhere in its statement α_2 is replaced by ω_2, let A be the set the set of grid points painted gray in Figure 1.5.2. Let $O = \{i, k, l, f\}$ (an ω_2-component of A), and let $Q = \{j\}$ (an ω_2-component of $\overline{A}$). Then $\partial(O, Q) = \{(i, j), (k, j)\}$ (see the heavily drawn edges in Figure 1.5.2). According to (i) of Claim 1.5.2, $k \in I$ and $j \in E$, but the ω_2-path $\langle k, m, c, j \rangle$ connecting k to j does not cross $\partial(O, Q)$. This contradicts (v) of Claim 1.5.2.

Where does this leave us? We need ω_2 (since it is used to describe the edges which make up the boundary), and we need an additional adjacency (for the validity of the rather desirable Claim 1.5.2). In subsequent chapters we show that there are grids for which we can get away with a single adjacency. However, since square grids are so ubiquitous, we will develop a theory capable of treating situations in which different adjacencies are used for sugar cubes and for their background.

1.6. Back to the Cuberille

In three-dimensional space, we have Fat Flies instead of Flat Flies. In Figure 1.6.1 we show one of these placed on a sugar cube.

What we would like to do now is to give an Algorithm for Fat Flies that has behavior resembling that of the Algorithm for Flat Flies as expressed in Claims 1.5.1 and 1.5.2. Before we do this, we have to resolve an ambiguity concerning the appropriate generalization of α_2 to the case of a cuberille.

One possible candidate, of course, is α_3. However, if we look at the geometrical interpretation of this adjacency, we see that it is the following. Two points of the cubic grid Z^3 are α_3-adjacent if, and only if, the corresponding Voronoi neighborhoods (i.e., voxels)

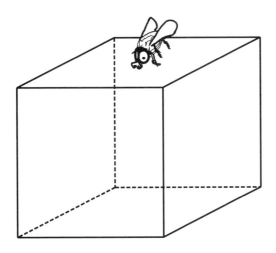

Figure 1.6.1. A Fat Fly on top of a single sugar cube.

have either exactly one face in common, or exactly one edge in common, or exactly one vertex in common. (Look at Figure 1.2.1: the grid points α_3-adjacent to the point g are exactly the other 26 points indicated in that figure.) The question is: for the purpose of the three-dimensional analog of Claim 1.5.1, do we want to consider two sugar cubes connected if they share only a vertex? This does not seem to be particularly desirable. As we have discussed in the previous section, such a connection is rather suspect. In the two-dimensional case, it was forced upon us to have an adjacency in addition to ω_2, since the very desirable Claim 1.5.2 does not remain true if we replace α_2 in it by ω_2. Now we give a trivial example to show that the situation is similar in three dimensions, i.e., if we replace Z^2 by Z^3 and both α_2 and ω_2 by ω_3 everywhere in Claim 1.5.2, then the resulting statement is false. The example is as follows.

Let $A = \{(1,0,0), (0,1,0), (0,0,1), (-1,0,0), (0,-1,0), (0,0,-1)\}$. (This can be thought of as meaning that A consists of the six cubes which surround the cube centered at the origin $(0,0,0)$.) Then $O = \{(1,0,0)\}$ is an ω_3-component of A and $Q = \{(0,0,0)\}$ is an ω_3-component of $\overline{A}$, such that $\partial(O,Q) = \{((1,0,0),(0,0,0))\}$ is not empty. By (i) of the new version of Claim 1.5.2, $(1,0,0) \in I$ and $(0,0,0) \in E$. However, the ω_3-path $\langle (1,0,0), (1,1,0), (0,1,0), (0,0,0) \rangle$ does not cross $\partial(O,Q)$, contradicting (v) of the new version of Claim 1.5.2.

If we want to have a cuberille analog of Claim 1.5.2, we need to replace Z^2 by Z^3. Also, geometric intuition tells us that we need ω_3, since it is used to describe the faces which make up the boundary. What we have just shown indicates that we need an additional adjacency, but does it have to be α_3?

Intuitively, there is an alternative three-dimensional generalization of α_2, one for which the corresponding voxels have either exactly one face in common or exactly one edge in common. Formally, for any positive integer N, we define the adjacency δ_N on Z^N as follows. For any c and d in Z^N,

$$(c,d) \in \delta_N \Leftrightarrow \left[(c,d) \in \alpha_N \text{ and } \sum_{n=1}^{N} |c_n - d_n| \leq 2 \right]. \tag{1.6.1}$$

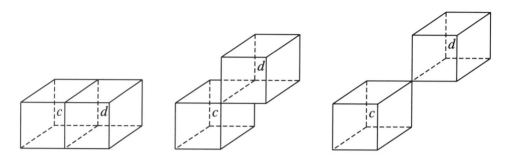

Figure 1.6.2. The grid points c and d on the left are α_3-, δ_3-, and ω_3-adjacent. The grid points c and d in the middle are α_3- and δ_3-adjacent but not ω_3-adjacent. The grid points c and d on the right are α_3-adjacent but are not δ_3-adjacent or ω_3-adjacent.

Clearly, $\delta_2 = \alpha_2$, and so δ_3 and α_3 are equally legitimate three-dimensional analogs of α_2. However, δ_3 is intuitively preferable over α_3, inasmuch as it does not allow adjacency by vertices only; see Figure 1.6.2. (For obvious reasons, ω_3 has been referred to both as face-adjacency and as 6-adjacency, δ_3 has been referred to both as face-or-edge-adjacency and as 18-adjacency, and α_3 has been referred to both as face-or-edge-or-vertex-adjacency and as 26-adjacency.) In view of this, now we can state our aim as follows.

Aim 1.6.1. *Design an Algorithm for Fat Flies that will have the following property. Suppose that in the cubic grid Z^3 (see Figure 1.2.1) the voxels of a finite number of grid points are filled with sugar cubes and that a Fat Fly is put on top of one of these sugar cubes into a voxel which is not occupied by a sugar cube (see Figure 1.6.1). Let g be the name of the grid point of the sugar cube on top of which the Fat Fly is placed into the voxel of a grid point named h. Let O be the δ_3-component of the set of grid points with sugar cubes which contains g and Q be the ω_3-component of the set of grid points without sugar cubes which contains h. After a finite number of steps of the Algorithm for Fat Flies, the Fat Fly should fly away and, at that time, the set of faces that have been dirtied should be exactly $\partial(O, Q)$.*

Before worrying about how to achieve this aim, let us discuss whether or not its output, namely $\partial(O, Q)$, is an intuitively desirable boundary. We say that it is, due to the truth of the following claim (whose proof is also postponed until we have reached the appropriate point in the development of our general theory).

Claim 1.6.2. *Let A be any nonempty proper subset of Z^3. Let O be a δ_3-component of A and Q be an ω_3-component of $\overline{A}$, such that $\partial(O, Q)$ is not empty. Then there exist two uniquely defined subsets I and E of Z^3 with the following properties.*
   (i) *$O \subset I$ and $Q \subset E$.*
   (ii) *$\partial(O, Q) = \partial(I, E)$.*
   (iii) *$I \cup E = Z^3$ and $I \cap E = \emptyset$.*
   (iv) *I is a δ_3-connected subset of Z^3 and E is an ω_3-connected subset of Z^3.*
   (v) *Every ω_3-path connecting an element of I to an element of E crosses $\partial(O, Q)$.*

Comparing Claims 1.5.2 and 1.6.2, we see that they are just two- and three-dimensional versions of each other. In fact, in our development, they will be trivial consequences of the same (much more general) theorem. This indicates the major motivation for what we are doing in this book: we wish to develop a general theory, the knowledge of which makes specific results (such as Claims 1.5.2 and 1.6.2) very easy to prove.

We also see that Claim 1.6.2 is a digital three-dimensional version of the Jordan Curve Theorem. Apart from it being a mathematically pleasing fact, this observation is of the utmost practical importance. In practical applications, we are interested in finding boundaries of objects for two reasons. One is to create computer graphic displays of them, such as the ones shown in Figure 1.1.3, and the other is to analyze certain properties of the objects, such as their volumes. From the graphical display point of view, property (v) can be used to prove that when we display the boundary $\partial(O, Q)$ as it appears from any exterior point, then the surface displayed will hide all the points interior to the object.

From the point of view of analysis, Claim 1.6.2 works as follows. Suppose that we have a device which has given us estimates of some physical quantity (such as the X-ray attenuation coefficient in a human body) at points of (a finite portion of) the cubic grid Z^3. Further suppose that we can identify from such data that set of grid points which are in cardiac muscle. Let A be the set of all such grid points. Assuming that A itself is δ_3-connected and that the set of grid points in any one of the chambers of the heart is an ω_3-component of $\overline{A}$, we can estimate the volume of the left ventricle, say, by selecting ω_3-adjacent voxels g and h, so that g is in the cardiac muscle and h is in the left ventricle, defining O and Q as in Aim 1.6.1, and estimating the volume of the left ventricle of the heart as the combined volume of the voxels associated with the grid points of the uniquely defined exterior E, whose existence is guaranteed by Claim 1.6.2. On the other hand, if we wish to find the volume of the whole heart (including the muscle and the chambers), then assuming that the set of grid points outside the heart form an ω_3-component of $\overline{A}$, we can select ω_3-adjacent voxels g and h, so that g is in the cardiac muscle and h is outside the heart, defining O and Q as in Aim 1.6.1, and estimating the volume of the whole heart as the combined volume of the voxels associated with the grid points of the uniquely defined interior I, whose existence is guaranteed by Claim 1.6.2. This discussion also indicates how the achievement of Aim 1.6.1 would allow us to detect individually the external and the internal boundaries of the cardiac muscle. Such considerations provide the basis of the practical importance of the theory developed in this book.

1.7. Algorithms for Fat Flies

In this section we discuss what can be done to adapt the Algorithm for Flat Flies to Fat Flies, so that we achieve Aim 1.6.1.

First we observe that just by simply changing the word "vertex" into the word "edge" we get a set of instructions that are meaningful in the three-dimensional case. However, the resulting algorithm will not fulfill Aim 1.6.1. Even if there is only one grid point with a sugar cube (see Figure 1.6.1), this simply modified version will fail. The Fat Fly will keep moving forward and, having dirtied only four faces of the cube, it will get back to the original face and will fly away. Clearly, $\partial(O, Q)$ in this case should consist of all six faces of the sugar

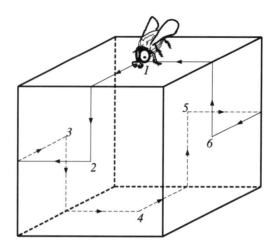

Figure 1.7.1. The route of a Model 1 Fat Fly on a single sugar cube. The numbers indicate the order in which the faces are visited.

cube, and the proposed algorithm resulted in only four of them being dirtied. So we need something more sophisticated. Consider the following.

Algorithm for Fat Flies Model 1

(1) Dirty the face on which you are standing.
(2) Crawl onto the face which meets the one on which you are standing at the edge in front of you.
(3) Turn right, and dirty the face on which you are standing.
(4) Crawl onto the face which meets the one on which you are standing at the edge in front of you.
(5) See if it is dirty.
    a.   If it is, fly away.
    b.   If it is not, turn left and start again at Instruction (1).

It turns out that this algorithm will behave as required for the case of a single sugar cube. It can be seen in Figure 1.7.1 that the Fat Fly under the control of the Model 1 algorithm will visit (and hence dirty) every face of a single sugar cube.

However, the very route of the Fat Fly as indicated in Figure 1.7.1 immediately supplies us with the hint that the Model 1 algorithm cannot behave as is desired according to Aim 1.6.1 because there are edges not traversed by this route. Now if we attach a second sugar cube to one of these edges, the grid points of the two sugar cubes are δ_3-adjacent, and so $\partial(O, Q)$ consists of the twelve faces of the pair of sugar cubes. Nevertheless, the route of the Fat Fly does not change; it still dirties only the same six faces. This is illustrated in Figure 1.7.2.

In fact, we see that this indicates a general principle. If the proposed algorithm is such that the edge toward which the Fat Fly starts crawling is not influenced by the absence or presence of nearby sugar cubes, then for the algorithm to work, it must traverse each edge when started on a face of a single sugar cube. (Otherwise, we can attach a second sugar

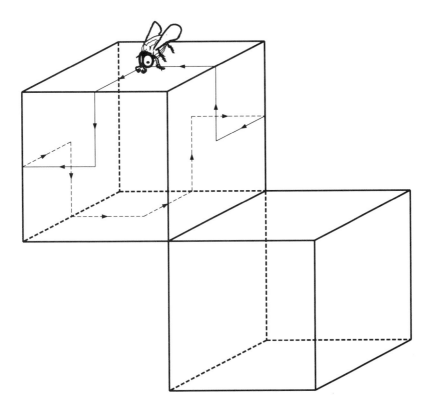

Figure 1.7.2. The route of a Model 1 Fat Fly on a pair of sugar cubes.

cube to the untraversed edge, just as indicated in Figure 1.7.2.) But if each edge is traversed once during a route on a single sugar cube, then each face must be visited twice. So the simple mechanism of dirtying the faces the first time they are visited and flying away the second time one is visited will not do. By itself this is a minor problem. We can have the fly "mark" a face the first time it visits it, "dirty" it when it finds that it has already been marked, and fly away when it finds that the face has already been dirtied. The more difficult question is: what should be the rule for deciding the edge towards which the Fat Fly starts crawling when it is on a particular face?

Now we establish a further general principle. Having accepted that the route of the Fat Fly must traverse each edge at least once, let us aim for the shortest route on a single sugar cube which satisfies this condition. If we draw the route of such a fly on the sugar cube as was done for Figure 1.7.1, then we see that on each face there have to be four arrows (one for each edge of the face), two of which point from the center of the face toward an edge and two which point from the other two edges toward the center of the face. Further, there has to be some consistency between these arrows: if an edge has an arrow pointing to it on one face, on the face which meets this edge the arrow has to be pointing away from the edge. Finally, if we also insist that the Algorithm for Fat Flies should be such that its local behavior is independent of its location (certainly a desirable property if we wish to design a simple algorithm), it follows that if an arrow on a face points away from an edge, then the

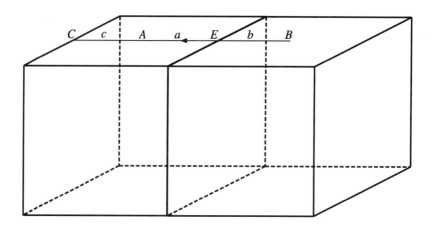

Figure 1.7.3. Illustration used in the discussion showing that the route of the Fat Fly cannot have opposing arrows meeting at the center of a face (i.e., an arrow from C to A is not allowed because of the arrow from E to A).

opposite edge on the same face must have an arrow pointing toward it. To see this, first let us make the requirement quite precise. What we would like to have is that the Algorithm for Fat Flies should be such that if the fly is started on any arrangement of sugar cubes and if we track its route until its flies away (by drawing arrows as described above), then on any face the arrangements of the arrows that is drawn should depend only on the orientation of that face. (There are six possible orientations, corresponding to the six faces of a single sugar cube.) Now consider the arrangement shown in Figure 1.7.3.

We are assuming here that there are no sugar cubes on top of the two indicated in the figure and that the route of the Fat Fly takes it from the edge point E to the face point A and hence that the arrow at a is as indicated. In this arrangement, the fly had to come from the face indicated by B and so the arrow at b has to point in the same direction as the arrow at a. By the assumption of the dependence of the arrows only on the orientation of the face, the arrow at c must point in the same direction as the arrow at b. It follows that if the arrow from the edge E points toward the center, then the arrow associated with the opposite edge C must point toward the edge.

This implies that the route of the Fat Fly for a single sugar cube must have the general form indicated by Figure 1.7.4. We say general form, rather than exact form, since the considerations above still leave us with some freedom. For example, the horizontal cycle of arrows around the vertical faces could be reversed without violating the conditions described above. The same is true for the two vertical cycles of arrows, leaving us with eight possible legitimate arrangements. However, these are symmetrical versions of each other and, for what we are going to do, there is no harm in restricting our attention to the arrangement shown in Figure 1.7.4. (This claim will be more rigorously justified later on.) A *legitimate route* for the Fat Fly is defined as one which follows the arrows and traverses each edge once.

In designing the Algorithm for Fat Flies, let us suppose that the fly is aware of the orientation of the face (one of six possible orientations) on which it finds itself. If it is now to behave so that it will follow one of the legitimate routes of Figure 1.7.4, then it has only two choices: go straight on or turn 90° in the legitimate direction. It is an interesting fact that

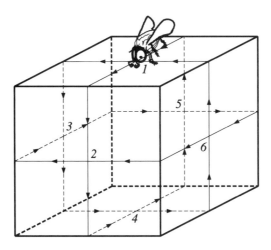

Figure 1.7.4. Legitimate routes of the Fat Fly on a single sugar cube follow the arrows and cross every edge once. The numbers refer to the orientations of the faces and not to the order in which they are visited.

the knowledge that the Fat Fly is visiting the face for the first or for the second time is not relevant to this decision: to achieve the required behavior, if it goes straight the first time, it must also go straight the second time and if it turns the first time, it must also turn the second time. (Otherwise, some edge would be traversed twice.) Thus the behavior of the Fat Fly can be controlled by a function g of the following kind. The domain of the function is $\{1, 2, 3, 4, 5, 6\}$, indicating one of the six possible orientations of a face of a sugar cube. (To be specific, consider Figure 1.7.1. Let 1 denote the orientation of the face with the Fat Fly on it, and let the orientations of the other faces be numbered by the order in which they are visited by the Fat Fly as it executes the Model 1 algorithm.) The value of the function g is either s (for go straight) or t (for turn). We call such a function a *routing function*. There are 64 possible routing functions. Now we are going to show that none of them do any good for us.

To prove such a bold statement, we need to make precise exactly what we mean by it. What we are going to do now is to introduce, for each routing function g, an Algorithm for Fat Flies and then show that none of them achieve what we wish to do.

Algorithm for Fat Flies Model g
(1) Mark the face on which you are standing.
(2) Crawl onto the face which meets the one on which you are standing at the edge in front of you. Check its orientation $i \in \{1, 2, 3, 4, 5, 6\}$.
(3) See if it is dirty.
    a.   If it is, fly away.
    b.   If it is not, see if it is marked.
        •   If it is, dirty it.
        •   If it is not, mark it.
(4) If $g(i) = t$, turn 90° in the legitimate direction indicated in Figure 1.7.4.
(5) Start again at Instruction (2).

Consider the particular case when $g(i) = t$, for $i \in \{2, 3, 4, 5, 6\}$. (We will decide on the value of $g(1)$ later on.) If the Fat Fly is placed on top of a single sugar cube as indicated in Figure 1.6.1, then after the sixth execution of Instruction (2) in the Algorithm for Fat Flies Model g it will have followed exactly the route indicated in Figure 1.7.1, and each of the faces have been marked, but none have been dirtied. At this moment the Fat Fly is not facing in the same direction as it was facing in its initial position indicated in Figure 1.7.1; rather, it is facing in the direction indicated by the incoming arrow on the top face. Now let us see what happens after this.

At this moment $i = 1$, and the face is marked but not dirty. According to Instruction (3) it gets dirtied. Now we have to apply Instruction (4).

If $g(1)$ happens to be t, then the Fat Fly turns to the left (see Figure 1.7.4), putting itself back into its initial position and repeats the same route as before until it gets back to face 1, which is dirty, and so the fly flies away. At this time exactly the same edges are covered as by the Model 1 algorithm, and so this behavior is not acceptable for the same reason as before (i.e., its output is incorrect for the situation presented in Figure 1.7.2).

On the other hand, if $g(1) = s$, then the Fat Fly will not turn and, according to Instruction (2), will climb onto the face with orientation $i = 3$. This is also marked but not dirty. According to Instruction (3) it gets dirtied and, according to Instruction (4), the Fat Fly turns to the right (see Figure 1.7.4). Then, according to Instruction (2), it climbs onto the face with orientation $i = 5$. This is also marked but not dirty. According to Instruction (3) it gets dirtied and, according to Instruction (4), the Fat Fly turns again to the right (see Figure 1.7.4). Then, according to Instruction (2), again it climbs onto the face with orientation $i = 1$. This face is dirty, and so the Fat Fly will fly away, having dirtied only three of the six faces of the sugar cube.

This shows that if $g(i) = t$, for $i \in \{2, 3, 4, 5, 6\}$, then the Algorithm for Fat Flies Model g will not do what we wish it to do (irrespective of the choice of $g(1)$). This behavior is not accidental, as we now show. (There is quite a bit of technical detail in the rest of this chapter. Those readers who are not particularly interested in mathematical details may wish to give it only a cursory reading.)

Let us call a routing function g *legitimate*, if whenever a Fat Fly is placed on a single sugar cube as indicated in Figure 1.6.1 and made to behave according to the Algorithm for Fat Flies Model g, by the time it flies away it will have traversed each edge at least once. (It is assumed in this definition that the Fat Fly will eventually fly away. This is all right, since in the execution of Instruction (3), if the Fat Fly does not fly away, it will either mark an unmarked face or it will dirty a marked face. As the algorithm keeps looping through Instructions (2)-(5), more and more faces get first marked and then dirtied and, since there are only six faces, eventually the face onto which the fly crawls has to be dirty.)

We have shown, before stating the definition, that g is not legitimate if $g(i) = t$, for $i \in \{2, 3, 4, 5, 6\}$, irrespective of the choice of $g(1)$. We could carry out a similar study for each of the remaining 62 routing functions to check whether or not it is legitimate. This could be done by simply following the behavior (under the control of the Algorithm for Fat Flies Model g) of a Fat Fly placed on a single sugar cube as indicated in Figure 1.7.1. However, doing this one by one for all the cases is tedious and one is prone to make an error. Here we suggest a more elegant alternative approach to get at the same information.

For now let us ignore the routing function itself and just look at the possible routes taken by a Fat Fly when placed on a single sugar cube as indicated in Figure 1.7.4 under the control of an Algorithm for Fat Flies Model g, for any g. The first stage of such a route is the same for all the gs: the Fat Fly moves from face *1* to face *2*. The simplification of the approach presented here relies on this fact: the initial segments of the routes taken by the Fat Fly are common for many different gs. So we list all possible initial routes of length one, two, three, and so on. To make sure that a route corresponds to a routing function, we must make sure that if a face is visited the second time along a route, then the route must be continued in a consistent way: if we went straight the first time, then we must do so the second time, and if we turned the first time, then we must turn the second time. A particular route need not be expanded any further if the face being visited is visited for the third time (this corresponds to the Fat Fly flying away in the algorithm). After we have traversed all twelve edges, this condition must be satisfied, and so the longest route that we consider will be of length 12. Also, if an initial route (of length no more than 12) is such that the same edge is visited twice (and hence there must be at least one untraversed edge at that time), then we need not expand that route any further, since from this time on the Fat Fly will keep retracing its previous route and so will never traverse the so far untraversed edges. (This situation corresponds to the case $g(1) = t$ in the example given above. The other reason for not continuing the route up to length 12 is the flying away before that, illustrated in the example above when $g(1) = s$.)

The initial routes of length up to five are all indicated in Figure 1.7.5. At the bottom of the figure, the * indicates that the particular route need not be expanded any further because the edge from the face with orientation *1* to the face with orientation *2* has been traversed twice. Since g is a function of the orientation of the face, the second time the face with orientation *2* is visited on a route, the same decision has to be taken regarding going straight as was taken the first time. This results in the arrow which ends at e not being one of a pair.

Now we can expand the routes which end their first five steps at the points indicated by the letters *a-l* in Figure 1.7.5. In Figure 1.7.6, we show all possible expansions from points *d*

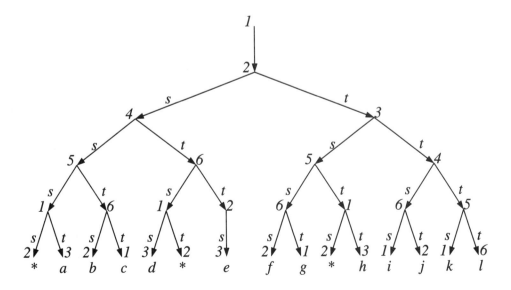

Figure 1.7.5. All routes up to length 5 that a Fat Fly may take under the control of an Algorithm for Fat Flies Model g.

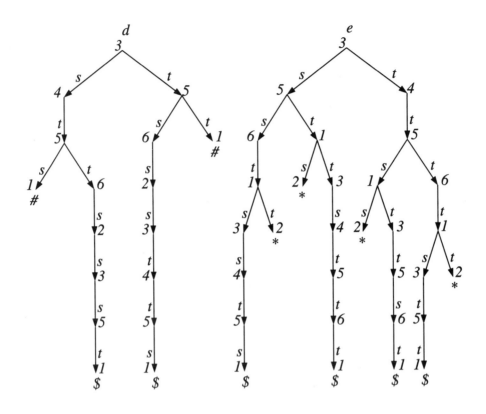

Figure 1.7.6. The rest of the possible routes that a Fat Fly may take under the control of an Algorithm for Fat Flies Model g expanded from points d and e in Figure 1.7.5.

and e. (We leave it to the reader to create similar representations of the possible expansions from the other ten points. It will turn out, in fact, from the mathematics to be discussed in the next section, that we do not actually need to do these other ten expansions to prove our results.) In Figure 1.7.6, just as in Figure 1.7.5, the * indicates that the particular route need not be expanded any further because the edge from the face with orientation 1 to the face with orientation 2 has been traversed twice. In Figure 1.7.6 we also see some # and $ signs. These mean that the route cannot be expanded further, since the face with orientation 1 has just been entered the third time and so the Fat Fly flies away at this point. The difference is that # indicates that at this time not all 12 edges have yet been traversed, whereas $ indicates the opposite. It is the latter which signals success: the routing function which gives rise to such a route is a legitimate one.

For example, the route which leads to the leftmost $ sign in Figure 1.7.6 shows that the routing function g for which $g(i) = s$, except for $i = 4$ or 5 is legitimate, and the route which leads to the rightmost $ sign in Figure 1.7.6 shows that the routing function g for which $g(i) = t$, except for $i = 1$ or 2 is legitimate. By expanding all the nodes, in this fashion we can discover all the legitimate routing functions. They are listed in Table 1.7.1. (The first and second row in this table, labeled by $d1$ and $d2$, respectively, correspond to the first and second $ signs in the expansion of the point d in Figure 1.7.6. Similarly, the rows labeled by

Table 1.7.1. Definitions of All the Legitimate Routing Functions

	1	2	3	4	5	6
d1	s	s	s	t	t	s
d2	s	s	t	t	s	s
i	s	t	t	s	s	s
f1	t	t	s	s	s	s
a1	t	s	s	s	s	t
c	s	s	s	s	t	t
e1	s	s	s	t	s	t
b2	s	s	t	s	t	s
f2	s	t	s	t	s	s
a2	t	s	t	s	s	s
g	s	t	s	s	s	t
b1	t	s	s	s	t	s
e2	t	s	s	t	t	t
e4	s	s	t	t	t	t
l	s	t	t	t	t	s
k	t	t	t	t	s	s
j1	t	t	t	s	s	t
h1	t	t	s	s	t	t
e3	t	s	t	t	s	t
j2	s	t	t	s	t	t
h2	t	t	s	t	t	s

e1, e2, e3, and *e4,* correspond to the four $ signs, left to right, respectively, in the expansion of the point *e* in Figure 1.7.6.)

The way this table has been produced is a fine example of the kind of thing that we wish to avoid. It is very specific to a particular set of definitions (a particular grid and particular adjacencies defined on it), and it is the result of an exhaustive investigation of a finite (but not small) number of cases. (The elementary, but undesirable, complete proof of the fact that Table 1.7.1 contains exactly the set of all legitimate routing functions requires producing five more figures of complexity similar to that of Figure 1.7.6. We did not include these since all the principles needed for producing them are already present in Figure 1.7.6 and, as we will see, it turns out that we do not actually need them.) One of the purposes of the general theory that follows is to provide the capability for creating proofs which have a large field of applicability and which are not dependent on investigating one by one a large number of specific arrangements. Nevertheless, we thought it worthwhile to show how Table 1.7.1 has been produced since (specifically) parts of it are needed for the proof of our next theorem and (generally) it indicates by its contrasting nature the desirability of the mathematically more elegant approaches of the later chapters.

To give a hint of what we judge to be mathematically elegant, we consider the following problem: suppose that we want to know, without necessarily listing them all, the number of legitimate routing functions. It turns out that this is a special case of a famous result of graph theory [12], and so we take a diversion into that discipline.

1.8. Digraphs

Formally, a *digraph* (sometimes called a directed graph) is an ordered pair (M, ρ), where M is any nonempty finite set (in the graph-theoretical context we refer to its elements as *nodes*) and ρ is an antireflexive binary relation on M, i.e., for any $c \in M$, $(c, c) \notin M$. (In the graph-theoretical context, we refer to the elements of ρ as *arcs*.) In view of this, all the terminology introduced in Section 1.4 is immediately applicable to digraphs. Since the ρ is fixed in a digraph, we use the phrases *adjacent to*, *adjacent from*, *path*, etc. instead of ρ-adjacent to, ρ-adjacent from, ρ-path, etc., respectively. For example, a path is a sequence $\langle c^{(0)}, \cdots, c^{(K)} \rangle$ of nodes such that, for $1 \le k \le K$, $\left(c^{(k-1)}, c^{(k)} \right)$ is an arc.

An example of a digraph is provided by Figure 1.7.4. In this case, $M = \{1, 2, 3, 4, 5, 6\}$, and $(i, j) \in \rho$ if, and only if, we can go in Figure 1.7.4 from the center of a face with orientation i to the center of a face with orientation j following two arrowed line segments across an edge. There is an alternative simple representation of this digraph, which is given in Figure 1.8.1. In this representation, the nodes correspond to the vertices of the hexagon, and the arcs are represented by the arrowed line segments connecting the vertices.

The first thing we do with this representation is to show that the eight possible ways of drawing the three cycles (one horizontal cycle of arrows and two vertical cycles of arrows, see Figure 1.7.4) on a sugar cube lead to exactly the same consequences. This is because we can label the orientations of the faces on the sugar cubes (after the cycles are drawn)

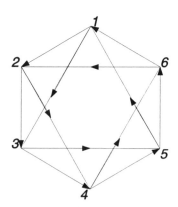

Figure 1.8.1. Alternative representation of the digraph of legitimate routes in Figure 1.7.4.

so that the resulting digraph has the alternative representation shown in Figure 1.8.1. (For example, we can call the orientation of the top face *1* and call the orientations of the two faces adjacent to it *2* and *3* so that *3* is adjacent to *2*, etc.) So by the simple trick of choosing the names of the orientations of the faces of a sugar cube in a way which depends on how the directions of the cycles have been selected, we end up with the same digraph in all cases. Hence, all the things that we will say regarding the specific case that we happened to select (the one in Figure 1.7.4) will also be valid for the seven alternative ways of directing the arrows around the sugar cube.

A path $\langle c^{(0)}, \cdots, c^{(K)} \rangle$ in a digraph (M, ρ) is said to be an *eulerian trail* if the following conditions are satisfied:

(i) $c^{(K)} = c^{(0)}$.

(ii) $\{c^{(0)}, \cdots, c^{(K)}\} = M$.

(iii) $\{(c^{(k-1)}, c^{(k)}) \mid 1 \le k \le K\} = \rho$.

(iv) For $1 \le k < k' \le K$, if $c^{(k)} = c^{(k')}$, then $c^{(k-1)} \ne c^{(k'-1)}$.

The first condition means that an eulerian trail has to be a *closed path* or, as we will also call such a thing, a *loop*. The second condition says that the eulerian trail has to visit all nodes. The third condition says that it makes use of all the arcs. The fourth condition says that no arc occurs in the trail more than once. For the digraph of Figure 1.7.4, the path $\langle 1, 2, 3, 4, 5, 6, 1 \rangle$ provided by the route of the Model 1 Fat Fly of Figure 1.7.1 satisfies Conditions (i), (ii) and (iv), but it fails to be an eulerian trail since it does not satisfy (iii); e.g., $(1, 3) \in \rho$, but 1 and 3 are never consecutive in the given path. The path $\langle 1, 3, 5, 1, 2, 4, 5, 1, 2, 3, 4, 6, 1 \rangle$ is also not an eulerian trail; it satisfies Conditions (i), (ii) and (iii), but fails to satisfy (iv) since, for example, *2* occurs at two different places in the path and is preceded by *1* in both cases. On the other hand, the path $\langle 1, 2, 4, 6, 1, 3, 4, 5, 6, 2, 3, 5, 1 \rangle$ provided by the route of the Model *d1* Fat Fly is an eulerian trail.

A little reflection will show that, in general, what we called legitimate routes for the Fat Fly in the previous section correspond exactly to those eulerian trails in the digraph of Figure 1.8.1 in which the first two elements of the path are *1* and *2*. The choice of these first two elements is not really important. In general, if $\langle c^{(0)}, \cdots, c^{(K)} \rangle$ is an eulerian trail and $1 \le k \le K$, then $\langle c^{(k)}, \cdots, c^{(K)}, c^{(1)}, \cdots, c^{(k)} \rangle$ is also an eulerian trail and, in fact, these two eulerian trails are considered one and the same. From the point of view of our specific example, consider any of the legitimate routing functions *g* of Table 1.7.1. Suppose that the Fat Fly under the control of the Algorithm for Fat Flies Model *g* (started on the top of the face with orientation *1* and looking towards the edge shared with the face with orientation *2*, see Figure 1.7.4) crawls during its execution from a face in orientation *i* to a face in orientation *j*. Now we are going to show that if the Fat Fly is placed on top of the face with orientation *i* and looking towards the edge of the face with orientation *j*, then it will still trace the same (in the sense of sameness just discussed) eulerian trail. (A particularly interesting special case is when the Fat Fly is placed on top of the face with orientation *1* and looking towards the edge shared with the face with orientation *3*. By the general result which we prove in the next paragraph, the same eulerian trail is traced as if we started the Fat Fly on top of the face with orientation *1* and looking towards the edge shared with the face with orientation *2*. Thus, if we have a legitimate routing function, the eulerian trail on a single sugar cube does not depend on the initial orientation of the Fat Fly.)

Clearly, if a Fat Fly is placed on top of the face with orientation *i* and looking towards the edge shared with the face with orientation *j* (call this Fly 2), then it will follow the end of the route of the Fat Fly that is placed on top of the face with orientation *1* and looking towards the edge shared with the face with orientation *2* (call this Fly 1), right until the moment at which that route returns to the face with orientation *1* for the second time and Fly 1 flies away. Unless $i = 1$ and $j = 2$, Fly 2 will not fly away at this time. If we could only be sure that now it will traverse the edge from the face with orientation *1* to the face with orientation *2*, then we would know that it would continue to follow the beginning of the route of Fly 1 and thus trace the same eulerian trail. However, this has to be the case, since Fly 1 is returning to the face with orientation *1* for the second time. The first time it returned to that face it came from the face with orientation either *5* or *6*, and so the second time it has to come from the other one (since the route is legitimate). Since at the first time Fly 1 returned to the face with orientation *1*, it had to turn to the face with orientation *3* (otherwise it would not have followed a legitimate route), when Fly 2 following the end of the route of Fly 1 arrives at the face with orientation *1* traversing the other edge, it must (under the control of the routing function g) now traverse the edge from the face with orientation *1* to the face with orientation *2*, as is required.

This discussion, combined with the third defining condition of an eulerian trail, means that if g is a legitimate routing function and the Fat Fly is placed on top of *any* face of a single sugar cube toward *either* of the edges indicated by an arrow in Figure 1.7.4, then under the control of the Algorithm for Fat Flies Model g it will visit each face twice and traverse each edge once before flying away.

Before discussing further what other ideas we can squeeze out of the classical notion of a digraph, let us return to the question that was asked at the end of the previous section: can we calculate the number of legitimate routing functions without having to explicitly list them? From the previous discussion we see that the number of legitimate routing functions is exactly the number of different eulerian trails of the digraph of Figure 1.8.1. There is a general formula applicable to *all* digraphs which have eulerian trails. (This formula is sometimes referred to as the BEST theorem, see [12], page 204.) Knowing this formula, we can calculate in about a minute (by working out the determinant of a simple 5×5 matrix) that the number of legitimate routing functions is indeed 21, as indicated in Table 1.7.1.

The next thing we are going to do is to show that even though there are 21 different eulerian trails of the digraph of Figure 1.8.1, there are only four such eulerian trails which are "essentially" different from each other, in the sense that all the others can be derived from these four by an easy manipulation.

Let us define the notation n_6 to mean exactly n, if $1 \le n \le 6$, with $7_6 = 1$ and $8_6 = 2$. Then we see that there are two types of arcs in the digraph of Figure 1.8.1: six of the form $(n, (n+1)_6)$ (let us call these *outer arcs*) and six of the form $(n, (n+2)_6)$ (let us call these *inner arcs*). As we describe a route of the Fat Fly in Figure 1.7.4 by a path in the digraph of Figure 1.8.1, we see that going straight in the former corresponds to switching from one type of arc to the other type in the latter and turning in the former corresponds to selecting the next arc to be of the same type as the previous one. For example, going straight in Figure 1.7.4 until we get back to the face in orientation *1* corresponds to the loop $\langle 1, 2, 4, 5, 1 \rangle$.

For any routing function g, the route of the fly under the control of the Algorithm for Fat Flies Model g (started in the situation shown in Figure 1.7.4) can be described by a path $\langle c^{(0)}, \cdots, c^{(K)} \rangle$, in which

(i) $c^{(0)} = 1$ and $c^{(1)} = 2$;

(ii) there exist k' and k'', such that $0 \le k' < k'' < K$ and $c^{(k')} = c^{(k'')} = c^{(K)}$;

(iii) there do not exist k, k' and k'', such that $0 \le k' < k'' < k < K$ and $c^{(k')} = c^{(k'')} = c^{(k)}$;

(iv) for $1 \le k \le K - 1$, $(c^{(k)}, c^{(k+1)})$ is an arc of the same type as $(c^{(k-1)}, c^{(k)})$, if $g(c^{(k)}) = t$ and is an arc whose type is not the same as that of $(c^{(k-1)}, c^{(k)})$, if $g(c^{(k)}) = s$.

As an example, for the legitimate routing function specified in the first row of Table 1.7.1, the route of the fly under the control of the Algorithm for Fat Flies Model *d1* (started in the situation shown in Figure 1.7.4) can be described by a path $\langle 1, 2, 4, 6, 1, 3, 4, 5, 6, 2, 3, 5, 1 \rangle$. This is indeed the route indicated in Figures 1.7.5 and 1.7.6 to the leftmost $ sign in Figure 1.7.6.

Now we make a very basic observation concerning Figure 1.8.1: if we rotate the figure clockwise 60° around its geometrical center, then all the nodes and arcs will map onto themselves (only the labels will appear to have been shifted). It follows that if a routing function g gives rise to an eulerian trail, then the routing function g', defined by $g'(n) = g((n+1)_6)$ must also give rise to an eulerian trail (and hence must be a legitimate routing function). To illustrate this, consider the just discussed eulerian trail $\langle 1, 2, 4, 6, 1, 3, 4, 5, 6, 2, 3, 5, 1 \rangle$, which is the route for the Algorithm for Fat Flies Model *d1* (started in the situation shown in Figure 1.7.4). Rotation clockwise by 60° turns this into the eulerian trail $\langle 6, 1, 3, 5, 6, 2, 3, 4, 5, 1, 2, 4, 6 \rangle$, which is the same eulerian trail as $\langle 1, 2, 4, 6, 1, 3, 5, 6, 2, 3, 4, 5, 1 \rangle$. In fact this is the route of the fly under the control of the Algorithm for Fat Flies Model *d2* (started in the situation shown in Figure 1.7.4), as indicated in Figures 1.7.5 and 1.7.6. That is, of course, what we expected, since simple observation of the first two lines of Table 1.7.1 shows that $d2 = d1'$.

Thus we have shown that the fact that *d2* is a legitimate routing function follows from the fact that *d1* is a legitimate routing function and does not require an independent confirmation of tracing the route of the fly under the control of the Algorithm for Fat Flies Model *d2*. By repeating this process, we find consecutively that i, $f1$, $a1$, c are all legitimate routing functions. In fact, by observing the entries of Table 1.7.1, we see that the facts that *d1*, *e1*, *e2*, and *e3* are also legitimate routing functions (explicitly demonstrated in Figures 1.7.5 and 1.7.6) imply that the other 17 entries are legitimate routing functions. Since we know from the BEST theorem that there are only 21 such functions, there is no need to look any further. So, as promised, we have shown that Table 1.7.1 is complete, without having had to produce expansions of the routes in addition to those shown in Figure 1.7.6.

Of equal importance, for what we want to do here, is that the same approach can be used to simplify proofs of negative results. For example, it is the case that if the Algorithm for Fat Flies Model *d1* does not satisfy Aim 1.6.1, then it follows that Algorithm for Fat Flies Model g will not satisfy Aim 1.6.1 if g is *d2*, i, $f1$, $a1$, or c. Similar statements can be made in place of *d1* for *e1*, *e2*, and *e3*. Now we demonstrate the truth of this claim in one case; it should be clear that all other cases follow by the demonstrated principle. What we are going to do is to show by a counterexample that the Algorithm for Fat Flies Model *d1* does not satisfy Aim 1.6.1 and then show that, from this counterexample, we can mechanically produce another counterexample which shows that the Algorithm for Fat Flies Model *d2* also does not satisfy Aim 1.6.1.

In Figure 1.8.2, we show a pair of sugar cubes, labeled *I* and *II*, respectively. The Fat Fly is placed on cube *I* on the face with orientation *1* (we use the notation *I1* to denote this face), looking towards the edge of face *I2*. Application of the Algorithm for Fat Flies Model *d1* provides us with the route $\langle I1, I2, I4, II4, II5, II6, II2, I2, I3, I5, I1, I2, \cdots \rangle$, which now

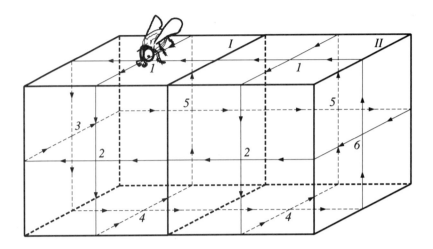

Figure 1.8.2. A pair of sugar cubes providing a counterexample which shows that the Algorithm for Fat Flies Model *d1* does not satisfy Aim 1.6.1.

repeats from the beginning, and so the Fat Fly will fly away without ever dirtying *III*. It follows therefore that the Algorithm for Fat Flies Model *d1* does not satisfy Aim 1.6.1.

This counterexample can be described as having been provided by placing the Fat Fly on a face with orientation *1* looking toward the edge of the face with orientation *2*, with a second sugar cube attached to the one on which the fly is placed at its face with orientation

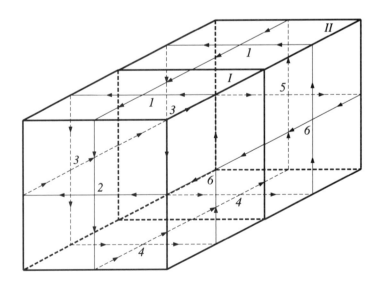

Figure 1.8.3. A pair of sugar cubes providing a counterexample which shows that the Algorithm for Fat Flies Model *d2* does not satisfy Aim 1.6.1.

6. Since $d2 = d1'$, to produce a counterexample for the Algorithm for Fat Flies Model $d2$, we need to place the Fat Fly on a face with orientation 6 looking toward the edge of the face with orientation 1, with a second sugar cube attached to the one on which the fly is placed at its face with orientation 5. The resulting sugar cubes have the arrangement shown in Figure 1.8.3, and it is easy to see that application of the Algorithm for Fat Flies Model $d2$ provides us with the route $\langle I6, I1, I3, II3, II4, II5, III1, I1, I2, I4, I6, I1, \cdots \rangle$, which now repeats from the beginning, and so the Fat Fly will fly away without ever dirtying $II6$. It follows therefore that the Algorithm for Fat Flies Model $d2$ also does not satisfy Aim 1.6.1.

1.9. So, What Can a Fat Fly Do?

What we would like is to have theorems for digital geometry, which have generality similar to the BEST theorem mentioned in the previous section and allow us to prove very easily whether or not there is an algorithm of a certain type (such as the Algorithm for Fat Flies Model g, for any g) which fulfills a particular set of requirements (such as those stated in Aim 1.6.1). Now we state and prove such a theorem. However, the proof at this early point in the book is not of the kind that meets our criterion of mathematical elegance.

Theorem 1.9.1. *For every routing function g, there exists a finite δ_3-connected set A of points in the cubic grid Z^3 such that $\overline{A}$ is ω_3-connected and, when the voxels associated with the grid points in A are filled with sugar cubes, one can place the Fat Fly on top of an uncovered face of a sugar cube so that if the Fat Fly is made to follow the instructions of the Algorithm for Fat Flies Model g, then it will fly away without having dirtied all the faces $\partial(A, \overline{A})$.*

Proof. First we observe that if g is not legitimate, then the theorem is valid. All we have to do is to let A give rise to two sugar cubes sharing only that edge of the one on which the Fat Fly is placed which would not be traversed by the time the Fat Fly would fly away if it were placed on that sugar cube on its own. Clearly, the presence of the second sugar cube does not alter the route of the Fat Fly and so none of the faces of the second sugar cube are ever visited, let alone dirtied. (This is just a generalization of Figure 1.7.2.)

Therefore need to worry only about the 21 legitimate routing functions given in Table 1.7.1. (All these will correctly detect the boundary when placed on top of a single sugar cube, as indicated in Figure 1.6.1.) In fact, in view of the discussion in the previous section, we need to prove the theorem only for the cases when g is $d1$, $e1$, $e2$, or $e3$. In fact, for the case when g is $d1$, we have already provided the proof using Figure 1.8.2.

The same two sugar cubes (indicated in Figure 1.8.2) also provide us with the proof when g is $e3$. We see that the Algorithm for Fat Flies Model $e3$, if started in the situation indicated in Figure 1.8.2, provides us with the route $\langle I1, I2, I4, II4, II5, III1, I1, I2 \cdots \rangle$, which now repeats from the beginning and so the Fat Fly will fly away without ever dirtying $I3$ (for example).

The same counterexample does not work for the cases when g is $e1$ or $e2$; instead we consider the arrangement of four sugar cubes (with a tunnel through the center of them) with

the Fat Fly placed in the position shown in Figure 1.9.1. The Algorithm for Fat Flies Model
$e1$ provides us with the route

$$\langle I1, I2, II1, II2, II4, II6, II2, II3, II5, II6, II1, II3, II4,$$
$$III2, III4, III6, III2, III3, III5, III6, III1, III3, III4, III5,$$
$$IV4, IV6, IV2, IV3, IV5, IV6, IV1, IV3, IV4, IV5, IV1, I5,$$
$$I1, I2, \cdots \rangle,$$

(1.9.1)

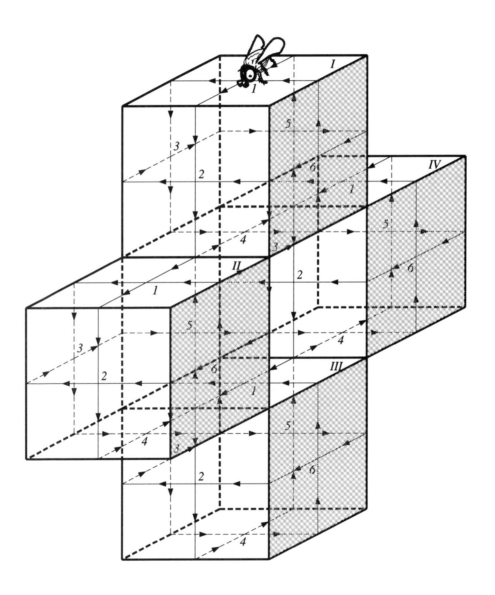

Figure 1.9.1. An arrangement of four sugar cubes (one face of each is painted gray to clarify the
location of a tunnel in the center of the four sugar cubes), providing a counterexample which shows
that the Algorithms for Fat Flies Models $e1$ and $e2$ do not satisfy Aim 1.6.1.

which now repeats from the beginning, and so the Fat Fly will fly away without ever dirtying *I3* (for example). Similarly, the Algorithm for Fat Flies Model *e2* provides us with the route

$$\langle I1, I2, II1, II3, II4, III2, III4, III6, III2, III3, III5,$$
$$IV4, IV6, IV2, IV3, IV5, IV1, IV3, IV4, IV5, IV6, IV1, I5, I6, \qquad (1.9.2)$$
$$I1, I2, \cdots \rangle,$$

which now repeats from the beginning, and so the Fat Fly will fly away without ever dirtying *I3* (for example).

For any routing function g, these constructions provide the finite δ_3-connected set A of points in the cubic grid Z^3 whose existence was claimed in the statement of the theorem. $\square$

The grand conclusion of all this is that there is no routing function g such that the Algorithm for Fat Flies Model g can serve as the algorithm of Aim 1.6.1. So we have to do something even more complicated; in the next section we present an example of an algorithm which satisfies the conditions of Aim 1.6.1.

1.10. Algorithms for Cloning Flies

One approach to achieving Aim 1.6.1 is due to Artzy [3]. His idea was to allow the flies to clone themselves and so be able to follow both directions on a face of a cube (see Figure 1.7.4) simultaneously. This results in the following algorithm.

Algorithm for Cloning Flies
(1) Crawl onto the face which meets the one on which you are standing at the edge in front of you.
(2) See if it is dirty.
    a.   If it is, fly away.
    b.   If it is not,
        •   dirty it,
        •   clone yourself into two flies facing in the two legitimate directions indicated in Figure 1.7.4 and both starting at Instruction (1).

If a Cloning Fly is placed on a single sugar cube, as shown in Figure 1.7.4, then after executing one iteration of Instructions (1) and (2) of the Algorithm for Cloning Flies, the situation will be as depicted in Figure 1.10.1.

Now let us see how the Algorithm for Cloning Flies behaves if the Cloning Fly is put on one of a pair of sugar cubes, as shown in Figure 1.8.2. The performance of the algorithm in this case is shown in Table 1.10.1. As can be seen from this table, after cycling six times through the two instructions of the Algorithm for Cloning Flies, all the flies will have flown away and the set of faces that have been dirtied by that time is exactly the boundary between the set of grid points of the two sugar cubes and the rest of the grid points.

Similarly, if we start the Algorithm for Cloning Flies on four sugar cubes, as shown in Figure 1.9.1, we also find that, after cycling eight times through the two instructions of the

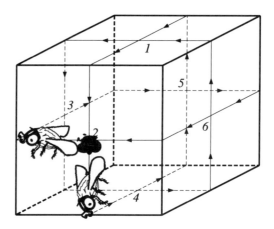

Figure 1.10.1. The situation after execution of one iteration of Instructions (1) and (2) of the Algorithm for Cloning Flies when the Cloning Fly is initially placed as depicted in Figure 1.7.4.

Algorithm for Cloning Flies, all the flies will have flown away, and the set of faces that have been dirtied by that time is exactly the boundary between the set of grid points of the four sugar cubes and the rest of the grid points.

Table 1.10.1. Performance of the Algorithm for Cloning Flies Initiated in the Situation Indicated in Figure 1.8.2

	Faces with Flies	Dirty Faces
Initially	I1	
After Instruction (1)	I2	
After Instruction (2)	I2	I2
After Instruction (1)	I3, I4	I2
After Instruction (2)	I3, I4	I2, I3, I4
After Instruction (1)	I4, I5, II4	I2, I3, I4
After Instruction (2)	I5, II4	I2, I3, I4, I5, II4
After Instruction (1)	I1, II5, II6	I2, I3, I4, I5, II4
After Instruction (2)	I1, II5, II6	I1, I2, I3, I4, I5, II4, II5, II6
After Instruction (1)	I2, I3, III1, III2, II6	I1, I2, I3, I4, I5, II4, II5, II6
After Instruction (2)	III1, II2	I1, I2, I3, I4, I5, III1, II2, II4, II5, II6
After Instruction (1)	I1, I2, II2, II4	I1, I2, I3, I4, I5, III1, II2, II4, II5, II6
After Instruction (2)		I1, I2, I3, I4, I5, III1, II2, II4, II5, II6

These two examples are special instances of a general phenomenon, since the Algorithm for Cloning Flies satisfies the conditions of Aim 1.6.1. To recapitulate what we mean by this, we state the following claim, whose complete proof is postponed till later.

Claim 1.10.1. *Suppose that in the cubic grid Z^3 the voxels of a finite number of grid points are filled with sugar cubes and a Cloning Fly is put on top of one of these sugar cubes into a voxel which is not occupied by a sugar cube. Let g be the name of the grid point of the sugar cube on top of which the Cloning Fly is placed into the voxel of a grid point named h. Let O be the δ_3-component of the set of grid points with sugar cubes which contains g and Q be the ω_3-component of the set of grid points without sugar cubes which contains h. After a finite number of steps of the Algorithm for Cloning Flies, all the flies will have flown away (due to Instruction (2)a), and, at that time, the set of faces that have been dirtied is exactly $\partial(O, Q)$.*

There is one aspect of this claim that we prove straight away because we need this partial result for proving a theorem later in this section.

Lemma 1.10.2. *Suppose that in the cubic grid Z^3 the voxels of a finite number of grid points are filled with sugar cubes and a Cloning Fly is put on top of one of these sugar cubes into a voxel which is not occupied by a sugar cube. After a finite number of steps of the Algorithm for Cloning Flies, all the flies will have flown away.*

Proof. Since only a finite number of grid points are filled with sugar cubes, there are only finitely many faces which are between a sugar cube and a voxel not occupied by a sugar cube. Therefore, Instruction (2)b can be executed only a finite number of times, since at the beginning there were only finitely many faces which may potentially be dirtied. After this last execution of Instruction (2)b, all remaining flies will execute Instructions (1) and (2)a, and so they will all fly away. □

Although the Cloning Flies are a wonderful concept, it is hard to manufacture them. Now we discuss an algorithm that can be run on a garden variety computer. Essentially the problem is that of simulating the behavior of a device which can follow many courses of action simultaneously (such as the Cloning Fly, if we consider it and all its descendants as one device) by a device which can execute only one action at a time. The usual trick is to create a *queue* of the things whose handling is postponed till the future. (According to the *Encyclopedia of Computer Science* [38] "Queues are nothing more than the waiting lines that have become an accepted and often frustrating fact of modern life." When used in an algorithm, queues are distinguished by the discipline according to which items are removed from it. For example, this can be first-in first-out or first-in last-out, usually abbreviated as FIFO and FILO, respectively [38]. The algorithms that we discuss in this book will achieve their aim for whatever queuing discipline is used. However, the computational costs may be quite dependent on the choice of the queuing discipline [10]. Since computational considerations are not at the essence of this book, we do not discuss this further and in all our examples will use the *first-in first-out queuing discipline*, according to which we remove the element which has been in the queue the longest.) In our simulation, the elements in the queue will be faces (which, in practice, will be identified by names, such as *I1*, *I2*, *III*, etc.; see Figures 1.8.2 and 1.9.1, Equations (1.9.1) and (1.9.2), and Table 1.10.1.) In creating a mental picture of what happens as a result of performing the instructions of the following algorithm, the reader should remember that the instructions are **not** given to a fly (of any of the kinds that we have discussed before) but to a computer, which is simulating in a sequential mode the behavior of the Cloning Fly, with the aim of eventually producing the same set of dirty faces as would be produced by the Cloning Fly.

Algorithm Simulating Cloning Flies

(1) Dirty the face which meets the one on which you are standing at the edge in front of you and put it into the queue.

(2) Remove a face f from the queue, and find the faces $f1$ and $f2$ in the boundary to which the Cloning Fly would get from the face f.

    a.   If $f1$ is not dirty, then dirty it and put it into the queue.

    b.   If $f2$ is not dirty, then dirty it and put it into the queue.

(3) Check if the queue is empty.

    a.   If it is, STOP.

    b.   If it is not, start again at Instruction (2).

In Table 1.10.2 we illustrate how the Algorithm Simulating Cloning Flies behaves if the Cloning Fly is put on one of a pair of sugar cubes, as shown in Figure 1.8.2. (To produce such a table, we have to decide on the order in which faces are removed from the queue in Instruction (2). Here we use the first-in first-out queuing discipline.) As can be seen from Table 1.10.2, after ten times cycling through Instructions (2) and (3) of the Algorithm Simulating Cloning Flies, the algorithm will STOP (since the queue is empty), and the set of faces that have been dirtied by that time is exactly the boundary between the set of grid points of the two sugar cubes and the rest of the grid points. Again, this example is a special instance

Table 1.10.2. Performance of the Algorithm Simulating Cloning Flies Initiated in the Situation Indicated in Figure 1.8.2, Using a First-In First-Out Queuing Discipline

	Faces in the Queue	Dirty Faces
After Instruction (1)	I2	I2
After Instruction (2)	I3, I4	I2, I3, I4
After Instruction (2)	I4, I5	I2, I3, I4, I5
After Instruction (2)	I5, II4	I2, I3, I4, I5, II4
After Instruction (2)	II4, I1, II5	I1, I2, I3, I4, I5, II4, II5
After Instruction (2)	I1, II5, II6	I1, I2, I3, I4, I5, II4, II5, II6
After Instruction (2)	II5, II6	I1, I2, I3, I4, I5, II4, II5, II6
After Instruction (2)	II6, II1	I1, I2, I3, I4, I5, II1, II4, II5, II6
After Instruction (2)	II1, II2	I1, I2, I3, I4, I5, II1, II2, II4, II5, II6
After Instruction (2)	II2	I1, I2, I3, I4, I5, II1, II2, II4, II5, II6
After Instruction (2)		I1, I2, I3, I4, I5, II1, II2, II4, II5, II6

of a general phenomenon, namely, that the Algorithm Simulating Cloning Flies satisfies the conditions of Aim 1.6.1 in the following sense.

Claim 1.10.3. *Suppose that in the cubic grid Z^3 the voxels of a finite number of grid points are filled with sugar cubes and a Cloning Fly is put on top of one of these sugar cubes into a voxel which is not occupied by a sugar cube. Let g be the name of the grid point of the sugar cube on top of which the Cloning Fly is placed into the voxel of a grid point named h. Let O be the δ_3-component of the set of grid points with sugar cubes which contains g and Q be the ω_3-component of the set of grid points without sugar cubes which contains h. After a finite number of steps of the Algorithm Simulating Cloning Flies, the algorithm will stop (due to Instruction (3)a) and, at that time, the set of faces that have been dirtied is exactly $\partial(O, Q)$.*

The complete proof of this claim is also postponed until later; here we restrict ourselves to showing that only one of Claims 1.10.1 and 1.10.3 need to be proved below, since both follow from the other. This theorem is preceded by two lemmas, the first of which proves the validity of a part of Claim 1.10.3.

Lemma 1.10.4. *Suppose that in the cubic grid Z^3 the voxels of a finite number of grid points are filled with sugar cubes and a Cloning Fly is put on top of one of these sugar cubes into a voxel which is not occupied by a sugar cube. The Algorithm Simulating Cloning Flies will stop after a finite number of steps.*

Proof. Since only a finite number of grid points are filled with sugar cubes, there are only finitely many faces which are between a sugar cube and a voxel not occupied by a sugar cube. This implies that after a finite number of steps in the execution of the Algorithm Simulating Cloning Flies, we will find that when executing Instruction (2) both $f1$ and $f2$ are dirty. Since the length of the queue is always finite and since any execution of Instruction (2) during which both $f1$ and $f2$ are found to be dirty reduces the length of the queue, the queue must eventually become empty, and the algorithm will stop. $\square$

Lemma 1.10.5. *Suppose that in the cubic grid Z^3 the voxels of some, but not of all, grid points are filled with sugar cubes and a Cloning Fly is put on top of one of these sugar cubes into a voxel which is not occupied by a sugar cube.*

   (i) *If any face gets dirtied by the Algorithm Simulating Cloning Flies, then that face will also get dirtied by the Algorithm for Cloning Flies.*

   (ii) *If it is the case that after a finite number of steps of the Algorithm for Cloning Flies all the flies will have flown away, then the faces which get dirtied by the Algorithm for Cloning Flies will also get dirtied by the Algorithm Simulating Cloning Flies.*

Proof. To show that (i) is true, we prove by induction that a face dirtied by the Algorithm Simulating Cloning Flies just before the nth execution of Instruction (2) also gets dirtied by the Algorithm for Cloning Flies. If $n = 1$, then the only face dirtied (due to Instruction (1)) by the Algorithm Simulating Cloning Flies is the one which meets the face on which the fly is originally placed at the edge in front of it. However, this face gets dirtied in the first execution of Instruction (2)b of the Algorithm for Cloning Flies. Now suppose that the induction hypothesis is true for some n, and consider a face f' dirtied by the Algorithm Simulating Cloning Flies just before the $(n + 1)$th execution of Instruction (2). If it had been already dirtied at the time just prior to the nth execution of Instruction (2), then (by the induction hypothesis) f' gets dirtied by the Algorithm for Cloning Flies. The only other possibility is that f' got dirtied during the nth execution of Instruction (2). This means that there must have been a face f which was in the queue just prior to the nth execution of Instruction (2), such that f' was one of the faces in the boundary to which the Cloning Fly would get from the face f. No face is ever put into the queue without being dirtied first and so f must have been dirtied prior to the nth execution of Instruction (2) and so (by the induction hypothesis) f gets dirtied by the Algorithm for Cloning Flies during an execution of its Instruction (2)b. Within the same instruction, the fly will clone itself into two flies facing in the two legitimate directions indicated in Figure 1.7.4 and both of these flies will (due to Instruction (1)) crawl onto the face which meets the one on which they are standing at the edge in front of them. One of these two faces will be f'. At this time, either f' is dirty or will get dirtied by the Algorithm for Cloning Flies in executing its Instruction (2)a. This completes the induction and hence the proof of (i).

Next we show that if after a finite number of steps of the Algorithm for Cloning Flies all the flies will have flown away, then it will also be the case that the Algorithm for Simulating Cloning Flies will stop after a finite number of steps. This is because, during the execution of Instruction (2) of the Algorithm for Simulating Cloning Flies, either the number of dirty faces increases or the length of the queue is reduced. Since, according to (i), the number of dirty faces cannot increase indefinitely, the length of the queue is reduced in all but finitely many executions of Instruction (2) of the Algorithm for Simulating Cloning Flies. Hence the queue must become eventually empty, and the algorithm must stop.

Assuming that after a finite number of steps of the Algorithm for Cloning Flies all the flies will have flown away, we now prove by induction the following statement: any face which gets dirtied during the nth execution of Instruction (2)b of the Algorithm for Cloning Flies, gets dirtied and is put into the queue at some time during the execution of the Algorithm Simulating Cloning Flies. If $n = 1$, then the only face that gets dirtied due to Instruction (2)b of the Algorithm for Cloning Flies is the one which meets the face on which the fly is originally placed at the edge in front of it. This face gets dirtied and is put into the queue by the execution of Instruction (1) of the Algorithm Simulating Cloning Flies. Now suppose that the induction hypothesis is true for some positive integer n and consider a face f' that gets dirtied during the $(n + 1)$th execution of Instruction (2)b of the Algorithm for Cloning Flies. Looking at the algorithm, we see that this could happen only if there is a face f such that f gets dirtied during the nth execution of Instruction (2)b of the Algorithm for Cloning Flies and f' is one of the faces in the boundary to which the Cloning Fly would get from the face f. By the induction hypothesis, f is put into the queue at some time during the execution of the Algorithm Simulating Cloning Flies. Since the queue is emptied in a finite number of steps, there will be an execution of Instruction (2) of the Algorithm for Simulating Cloning Flies which removes the face f from the queue and (if f' is not dirty already, in which case it must have been dirtied and put into the queue previously) it dirties f' and puts it into the queue. This completes the induction and hence the proof of (ii). $\square$

Theorem 1.10.6. *Suppose that in the cubic grid Z^3 the voxels of a finite number of grid points are filled with sugar cubes and a Cloning Fly is put on top of one of these sugar cubes into a voxel not occupied by a sugar cube. Let g be the name of the grid point of the sugar cube on top of which the Cloning Fly is placed into the voxel of a grid point named h. Let O be the δ_3-component of the set of grid points with sugar cubes which contains g and Q be* *the ω_3-component of the set of grid points without sugar cubes which contains h. Then the following two statements are equivalent.*

(i) *After a finite number of steps of the Algorithm for Cloning Flies, all the flies will have flown away (due to Instruction (2)a) and, at that time, the set of faces dirtied is exactly $\partial(O, Q)$.*

(ii) *After a finite number of steps of the Algorithm Simulating Cloning Flies, the algorithm will stop (due to Instruction (3)a) and, at that time, the set of faces dirtied is exactly $\partial(O, Q)$.*

Proof. We know that after a finite number of steps of the Algorithm for Cloning Flies all the flies will have flown away (Lemma 1.10.2) and that after a finite number of steps of the Algorithm Simulating Cloning Flies the algorithm will stop (Lemma 1.10.4). We also know that any face gets dirtied by one of the algorithms if, and only if, it gets dirtied by the other one (Lemma 1.10.5). □

1.11. An Efficient Implementation

Generally speaking, the details of computer implementation of the algorithms proposed in this book will not be of one of our concerns. Rather, we will be concerned with the input-output properties of the algorithms: what will they produce when started in some particular situation? (Examples of such properties are given in Claims 1.5.1, 1.10.1, 1.10.3, and Theorem 1.9.1.) Usually, we let the reader find out about details of computer implementation from the literature; in the case of the Algorithm for Simulating Cloning Flies, such details can be found in [10]. However, there is one implementational aspect of this algorithm which is of essential mathematical interest. In fact, it is one of the most pleasing aspects of Artzy's original idea [3].

The potential difficulty that arises is subtle. It comes from the phrases "If $f1$ is not dirty , ..." and "If $f2$ is not dirty , ..." at the beginning of Instructions (2)a and (2)b for the Algorithm Simulating Cloning Flies. How do we find out if a face is dirty? This was easy to do while we were producing Table 1.10.2, because the number of dirty faces never got beyond ten. However, in a medical application, surfaces such as the ones depicted in Figure 1.1.3 may consist of millions of faces. In such a situation, checking whether or not a face has already been dirtied requires a great deal of computational resources (meaning storage space and/or computer time).

The observation that allows us to considerably reduce the size of this problem is the following: there are exactly two ways to get to any face in the boundary. Now we rephrase this observation in standard mathematical terminology. Consider the digraph whose nodes are the faces in the boundary and in which a face f is adjacent to a face f' if, and only if, f' is one of the faces in the boundary to which the Cloning Fly would get from the face f. Then we see that every face in the boundary is adjacent from exactly two faces in the boundary; in other words, every node in the digraph has *indegree* two.

We use this mathematical property by introducing the additional concept of a *list* of faces into the algorithm. The basic idea will be that we can keep the number of faces in the list generally much smaller than the number of dirty faces, but the algorithm can do everything that it must do by checking for particular faces whether or not they are in the list, rather than whether or not they are dirty.

Artzy's Algorithm

(1) Dirty the face which meets the one on which you are standing at the edge in front of you, put it into the queue and put two copies of it in the list.

(2) Remove a face f from the queue, and find the faces $f1$ and $f2$ in the boundary to which the Cloning Fly would get from the face f.

    a.   Try to find one copy of $f1$ in the list.

       •   If successful, remove it from the list.

       •   If not, then dirty $f1$, and put it into the queue and the list.

    b.   Try to find one copy of $f2$ in the list.

       •   If successful, remove it from the list.

       •   If not, then dirty $f2$, and put it into the queue and the list.

(3) Check if the queue is empty.

    a.   If it is, STOP.

    b.   If it is not, start again at Instruction (2).

In Table 1.11.1 we illustrate how Artzy's Algorithm behaves (using a first-in first-out queuing discipline) if the Cloning Fly is put on one of a pair of sugar cubes, as shown in Figure 1.8.2. As can be seen from this table, after cycling ten times through Instructions (2) and (3) of Artzy's Algorithm, the algorithm will STOP (since the queue is empty) and the set of faces that have been dirtied by that time is exactly the boundary between the set of grid points of the two sugar cubes and the rest of the grid points. In fact, by comparing

Table 1.11.1. Performance of Artzy's Algorithm Initiated in the Situation Indicated in Figure 1.8.2, Using a First-In First-Out Queuing Discipline

	Queue	List	Dirty Faces
After (1)	I2	I2, I2	I2
After (2)	I3, I4	I2, I2, I3, I4	I2, I3, I4
After (2)	I4, I5	I2, I2, I3, I5	I2, I3, I4, I5
After (2)	I5, II4	I2, I2, I3, II4	I2, I3, I4, I5, II4
After (2)	II4, I1, II5	I1, I2, I2, I3, II4, II5	I1, I2, I3, I4, I5, II4, II5
After (2)	I1, II5, II6	I1, I2, I2, I3, II4, II6	I1, I2, I3, I4, I5, II4, II5, II6
After (2)	II5, II6	I1, I2, II4, II6	I1, I2, I3, I4, I5, II4, II5, II6
After (2)	II6, II1	I1, I2, II1, II4	I1, I2, I3, I4, I5, II1, II4, II5, II6
After (2)	II1, II2	I1, I2, II2, II4	I1, I2, I3, I4, I5, II1, II2, II4, II5, II6
After (2)	II2	I2, II4	I1, I2, I3, I4, I5, II1, II2, II4, II5, II6
After (2)			I1, I2, I3, I4, I5, II1, II2, II4, II5, II6

Table 1.10.2 and Table 1.11.1, we see that the behavior of the Algorithm Simulating Cloning Flies and of Artzy's Algorithm is the same, except that the latter has the additional list of faces. Again, these statements are special instances of some general phenomena, which we now state. The proofs are postponed till later. The claims stated here will be corollaries of more general results derived in last chapter of the book. Our aim between here and there is the production of a geometry applicable to digital spaces. The fact that claims such as the two that follow are immediate consequences of some general theorems within the geometry developed in the following chapters confirms the successful achievement of this aim.

Claim 1.11.1. *Suppose that in the cubic grid Z^3 the voxels of some, but not of all, grid points are filled with sugar cubes and that we put a Cloning Fly on top of one of these sugar cubes into a voxel not occupied by a sugar cube and then run both the Algorithm for Simulating Cloning Flies and Artzy's Algorithm. For any positive integer n, either both algorithms will have stopped in less than n cycles through their respective Instructions (2) and (3) or at both just before and just after the nth execution of Instruction (2) the queue and the set of dirty faces is the same for both algorithms.*

Claim 1.11.2. *Suppose that in the cubic grid Z^3 the voxels of a finite number of grid points are filled with sugar cubes and a Cloning Fly is put on top of one of these sugar cubes into a voxel not occupied by a sugar cube. Let g be the name of the grid point of the sugar cube on top of which the Cloning Fly is placed into the voxel of a grid point named h. Let O be the δ_3-component of the set of grid points with sugar cubes which contains g and Q be the ω_3-component of the set of grid points without sugar cubes which contains h. Artzy's Algorithm will stop (due to Instruction (3)a) after a finite number of steps and, at that time, the set of faces dirtied is exactly $\partial(O, Q)$.*

We complete this chapter with a theorem which makes precise the limitation on the size of the list during the execution of Artzy's Algorithm. This limitation is what makes deciding whether or not a face is in the list in Artzy's Algorithm much less demanding on computational resources than deciding whether or not a face is dirty in the Algorithm for Simulating Cloning Flies.

Theorem 1.11.3. *Just before and just after the execution of Instruction (2) in Artzy's Algorithm, the list is exactly twice as long as the queue.*

Proof. This is trivially valid just before the first execution of Instruction (2). Now assume that the list is twice as long as the queue just before some execution of Instruction (2). Since neither the list nor the queue is changed during the execution of Instruction (3), it is sufficient to prove that the list will be twice as long as the queue after the execution of Instruction (2). At the beginning of the execution of Instruction (2), the length of the queue is reduced by one. Now we consider three cases.

If before the execution of Instruction (2) both *f1* and *f2* are in the list, then they are both removed from the list and so the length of the list is decreased by two. If before the execution of Instruction (2) neither *f1* and *f2* are in the list, then they are both added to the queue and the list, resulting in a total increase of one in the length of the queue and a total increase of two in the length of the list. If before the execution of Instruction (2) exactly one of *f1* and *f2* are in the list, then that one gets added to the list and the queue, whereas the other is removed from the list, resulting in no change in the length of the queue and the list during the execution of Instruction (2). □

1.12. Exercises

1.1.   For any positive integer N, any positive real number r and for any $c \in R^N$, we define (as usual) the N-dimensional *ball* of radius r centered at c as

$$B_{r,c} = \left\{ v \in R^N \mid \|v - c\| \leq r \right\}. \tag{1.12.1}$$

Prove that if G is a (D, d) grid in R^N and $g \in G$, then $B_{d,g} \subset N_G(g) \subset B_{D,g}$.

1.2.   Prove that, for every positive integer N and positive real number δ, δZ^N is a Delone grid.

1.3.   Prove that if G is a Delone grid and C is a nonnegative integer, then the set $\{ g \in G \mid \|g\| \geq C \}$ is infinite.

1.4.   As usual, we call a subset M of R^N a *bounded set* if there exists a real number B such that, for all $v \in M$, $\|v\| \leq B$. Let G be a Delone grid in R^N and let F be a subset of G. Prove that F is finite if, and only if, $\bigcup_{g \in F} N_G(g)$ is bounded.

1.5.   For any Delone grid G in R^2, prove that:
 (i) The Voronoi neighborhood in G of any element g of G consists of a polygon and its interior. (Recall that a *polygon* is a planar figure consisting of $p \geq 3$ points, each of which is called a *vertex* of the polygon, and p line segments, each one of which is called an *edge* or side of the polygon, whose end points are the vertices and which have no elements in common except their end points.)
 (ii) If the Voronoi neighborhoods in G of any two distinct elements g and h of G have a point in common, then either that point is a vertex of both of the polygons or the two polygons share an edge containing that point.

1.6.   Discuss the analog of the previous exercise in R^3.

1.7.   On the grid Z^4 in R^4 define the binary relation ϕ by

$$(c, d) \in \phi \Leftrightarrow \left[(c, d) \in \alpha_4 \text{ and } \left(c_4 \neq d_4 \text{ or } \sum_{n=1}^{3} |c_n - d_n| \neq 2 \right) \right]. \tag{1.12.2}$$

How many grid points are ϕ-adjacent to $(0, 0, 0, 0)$?

1.8.   In the proof of Theorem 1.9.1 we have used two different arrangements of sugar cubes (shown in Figures 1.8.2 and 1.9.1) to provide our counterexamples. Is there a single arrangement of sugar cubes which will provide a counterexample to all four cases?

1.9.   Describe the performance of Artzy's Algorithm initiated in the situation indicated in Figure 1.9.1, using a first-in first-out queuing discipline.

1.10.   Show that whenever Artzy's Algorithm stops, the list is empty at that time.

2
Enhancing the Cube

"The fabled El Corazón arced high in the air, refracting sparkling emerald shafts of light."

J. Wilder, *Romancing the Stone*, Chapter Twenty-Nine.

2.1. Why Study Noncubic Grids?

The cubic grids in R^3 (just as the square grids in R^2) are ubiquitous in image processing. In fact, the percentage of publications which use grids other than these is very small. Even in mathematics, one can develop a full-fledged theory of topology (see Chapter 4) based only on these grids; see, e.g., [33]. Therefore it is reasonable to ask why we should bother to develop a theory more general than the one based on the cubic grids in R^3 and on the square grids in R^2. In this chapter we respond to this question. The current section compares the cubic grids with the so-called fcc grids and shows that the latter have several advantages. In the next section we discuss a host of different spaces and adjacencies which have arisen for various reasons in the literature.

The cubic grid δZ^3 ($\delta > 0$) consists of all triples $(\delta c_1, \delta c_2, \delta c_3)$, where c_1, c_2, c_3 are integers. The associated voxels are cubes of volume δ^3. In computer graphic displays of surfaces consisting of faces of such cubes, one can make good use of the facts that all the faces have the same shape, at least half of them are automatically hidden (and one can easily predetermine which half), the visible faces lie in one of at most three possible orientations, and one can Z-buffer them very efficiently. (The verb "to Z-buffer" refers to one of the approaches to ensuring that only the visible parts of the surface contribute to the final computer graphic display.) Since computer graphics is not the main subject matter of this book, we leave it to the reader to find a more detailed description in the literature of the Z-buffer algorithm and the justification of the claims we have just made regarding computer graphic displays of surfaces based on the cubic grid (see, for example, Sections 15.3 and 15.4 of [13]). Here we discuss only two disadvantages of the cubic grids for computer graphic display of natural objects.

We have already touched upon the first disadvantage in the previous chapter. We have pointed out that the important Claim 1.6.2 cannot be true without using an adjacency in addition to ω_3. In that claim we have used δ_3 for A and ω_3 for its complement $\overline{A}$. However, we will see (from the general theory to be developed below) that the claim is equally valid if we use ω_3 for A and δ_3 for its complement $\overline{A}$. Now, if we think of A as the set of grid points occupied by sugar cubes, we see that there is an undesirable arbitrariness here: we cannot use ω_3 for both A and its complement, but we may choose to use δ_3 for either to make the claim true. Hence the decision as to where to use δ_3 is arbitrary from the mathematical point of view. Our image of a fly crawling on the surface made it more natural to use δ_3 for A (otherwise the fly has to squeeze itself between the touching edges of two sugar cubes), but surely this pictorial representation is hardly a firm guiding principle as to what is the appropriate choice in a real-life application area. Also, as we have pointed out, our choice allows rather flimsily connected objects, whose connectedness is suspect from a physical point of view. For example, if A gives rise to a pair sugar cubes, as shown in Figure 1.7.2, then A is a δ_3-component, $\overline{A}$ is an ω_3-component, and $\partial(A, \overline{A})$ consists of the twelve faces of the two sugar cubes. It is easy to see that in this case the interior I and the exterior E of Claim 1.6.2 are in fact A and $\overline{A}$, respectively. As asserted in Claim 1.6.2, I is δ_3-connected. However, if we define the *topological interior* of a set H of points in R^3 as the set of all those points v, for which there exists a positive real number r such that $B_{r,v} \subset H$ (see page 42), then we see that the topological interior of the union of the Voronoi neighborhoods of the points in A (i.e., of the two sugar cubes) is not a connected set in the topological sense. (We will give a precise definition of topological connectedness in Chapter 4. For now we may intuitively think of the lack of topological connectedness of the interiors of the two sugar cubes as follows: there is no continuous curve which starts in the interior of one of the sugar cubes and ends in the interior of the other which does not leave the union of the two interiors somewhere along its route. This is because the common edge of the two sugar cubes is clearly not in the interior of their union.) This lack of correspondence between connectedness of the interiors as used in Claim 1.6.2 and the topological connectedness of the corresponding topological interiors in R^3 is just a more formal mathematical demonstration of the unease we have already expressed in the previous chapter: is it (biologically, physically, whateverly) reasonable to consider the pair of sugar cubes in Figure 1.7.2 to be connected to each other?

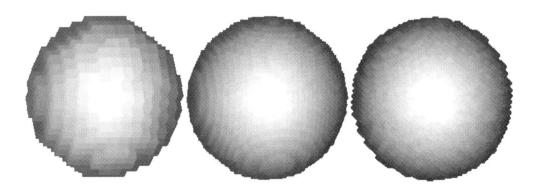

Figure 2.1.1. Computer graphic display of a sphere based on different grids. Left: the cubic grid $2Z^3$. Middle: the cubic grid Z^3 (i.e., the volume of the voxels is one eighth of that in the cubic grid $2Z^3$). Right: the fcc grid $F_{\sqrt[3]{4}}$ (i.e., the same volume voxels as in the cubic grid $2Z^3$). In all cases the computer graphic display is produced on the square grid $\frac{1}{8}Z^2$ using standard techniques, such as those described, for example, in Chapter 15 of [13].

The second disadvantage of using a cubic grid is the following. Since faces adjacent at edges make right angles with each other, the computer graphic display of the surface may appear blocky. This can be overcome by a number of techniques (for example, by using smaller voxels), but all of them result in an increase in computational complexity compared to just using the original voxel faces. Now we illustrate this on a simple example.

Consider all the grid points of a grid $G \subset R^3$ which are inside a centrally located sphere of radius 15; i.e., let $A = \{ g \in G \,|\, \|g\| < 15 \}$. For any reasonable adjacencies for a reasonable grid G, both A and $\overline{A}$ will be components and $\partial(A, \overline{A})$ can be considered a digital approximation of the spherical surface. In Figure 2.1.1 we illustrate computer graphic displays of such digital approximations for different grids. (A standard methodology based on the one described in Chapter 15 of [13] was used to create these displays. In all cases, the surface is first tilted by 30° about the horizontal axis parallel to the display screen and then rotated by 15° around the vertical axis parallel to the display screen.) It can be seen that the appearance of the display based on an fcc grid (defined below) is much smoother than that based on the cubic grid using voxels of the same volume and is comparable to that based on the cubic grid requiring eight times as many voxels per unit volume.

Now we discuss a family of grids which provides an improvement over the cubic grids from the point of view of these two disadvantages. For any positive real number ϕ, we define a *face-centered cubic grid* (or *fcc grid*, for short) as the set of points

$$F_\phi = \{(\phi c_1, \phi c_2, \phi c_3) \mid c_1, c_2, c_3 \in Z \text{ and } c_1 + c_2 + c_3 \equiv 0 \pmod 2\}. \qquad (2.1.1)$$

Each of the associated voxels is a *rhombic dodecahedron* (i.e., a twelve-faced solid, for which every face is an identical rhombus), as can be seen from Figure 2.1.2. We will sometimes abbreviate F_1 as F.

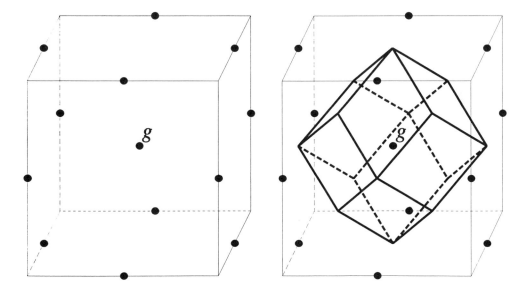

Figure 2.1.2. On the left we show a point g of the fcc grid $F = F_1$ together with the 12 grid points which lie on the $2\times2\times2$ cube whose center is g. On the right we show (using heavy lines) the Voronoi neighborhood of g (a voxel, whose shape is a rhombic dodecahedron).

Figure 2.1.3. A rhombic-dodecahedral-shaped crystal consisting of hexons. (A hexon is a major coat protein of the human adenovirus.) Four hexon molecules form a small cube with edge 150.7 Å that is the repeating unit of the crystal. The approximately 0.6 mm crystal shown contains 40,000 repeating units on each edge and a total of approximately 10^{13} hexons. X-ray crystallography has been used to determine the three-dimensional structure of a hexon by analyzing its diffraction pattern [4]. (Picture by courtesy of Roger M. Burnett of The Wistar Institute, Philadelphia, PA.)

The rhombic dodecahedron is not simply a figment of mathematical imagination invented for the purposes discussed in this chapter. It is a shape which in fact appears in nature, as illustrated in Figure 2.1.3.

To see how the fcc grids avoid the previously stated (and other) disadvantages of cubic grids, now we define a notion of adjacency (i.e., a symmetric binary relation) for the fcc grid. For any pair (c, d) of grid points in F_ϕ,

$$(c, d) \in \beta_\phi \Leftrightarrow \|c - d\| = \phi\sqrt{2} \, . \tag{2.1.2}$$

Obviously, for any grid point in F_ϕ, there are twelve grid points in F_ϕ which are β_ϕ-adjacent to (and from) it. These are illustrated, for the case $\phi = 1$, in Figure 2.1.2. An alternative illustration of the twelve grid points which are adjacent in the fcc grid to a given grid point is provided in Figure 2.1.4, a photograph of a work of the artist Brooks Betts, who had not heard of fcc grids at the time it was created. Consider the "egghead" in Figure 2.1.4 as the given grid point. The twelve grid points adjacent to it are represented by the twelve screw eyes, each at the center of an edge of the cubical frame. The nylon filaments connecting the egghead to the screw eyes may be interpreted as representing the adjacencies. This is indeed a face-centered cubic grid!

As illustrated in Figure 2.1.5, two grid points in F_ϕ are β_ϕ-adjacent if, and only if, the voxels associated with them share exactly one face. In this sense, β_ϕ-adjacency for an fcc grid is similar to the adjacency ω_3 for the cubic grid Z^3. However, for the fcc grid we have the following version of Claim 1.6.2, which is stronger than Claim 1.6.2 in the sense that it does not need to use an adjacency in addition to β_ϕ. (In stating this claim here, we are jumping the gun a bit. The statement of the claim uses concepts which were defined in the previous chapter for the adjacencies ω_N, such as the boundary between two subsets of the set of all grid points. Strictly speaking, we should redefine all these concepts with Z^N replaced by F_ϕ and ω_N replaced by β_ϕ at every place where they occur in the definitions. We do not do this here for two reasons: first, it should be pretty obvious to the reader what needs to be done and, second, the more general definitions of the next chapter will take care of this temporary lack of mathematical rigor.)

Figure 2.1.4. EGGHEAD. Wood, plaster, nylon filament, screw eyes, nails. $10.5'' \times 10'' \times 10''$. ©Brooks Betts 1973.

Claim 2.1.1. *Let A be any nonempty proper subset of F_ϕ (for some $\phi > 0$). Let O be a β_ϕ-component of A and Q be a β_ϕ-component of $\overline{A}$, such that $\partial(O, Q)$ is not empty. Then there exist two uniquely defined subsets I and E of F_ϕ, which have the following properties.*

(i) *$O \subset I$ and $Q \subset E$.*
(ii) *$\partial(O, Q) = \partial(I, E)$.*
(iii) *$I \cup E = F_\phi$ and $I \cap E = \emptyset$.*
(iv) *Both I and E are β_ϕ-connected subsets of F_ϕ.*
(v) *Every β_ϕ-path connecting an element of I to an element of E crosses $\partial(O, Q)$.*

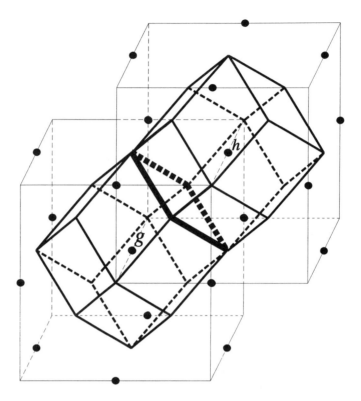

Figure 2.1.5. Two grid points g and h in the fcc grid F_ϕ are β_ϕ-adjacent if, and only if, the voxels corresponding to them share exactly one face. (Therefore it follows that the topological interior of the union of voxels of adjacent grid points is topologically connected.) Every edge is shared by three voxels, and so the normals between two faces meeting at an edge are at 60° to each other.

This result also will be seen to be the consequence of more general results later on. (In fact, part of the motivation of the abstract theory that we will develop is that Claims 1.5.2, 1.6.2, and 2.1.1 are all corollaries of the same theorem.) For now observe only that the first of the above mentioned disadvantages of a cubic grid disappears when we use an fcc grid: in Claim 2.1.1 we use only one adjacency (consequently, the objectionable arbitrariness as to which adjacency to use for A and which for $\overline{A}$ disappears) and this adjacency is "solid," in the sense that if two voxels correspond to adjacent grid points, then they must share a whole face in common. (In fact, we see that it is not possible to define a face-or-edge-adjacency which is not also a face-adjacency: whenever two of the rhombic dodecahedra share an edge, they must also share a whole face. Thus the β_ϕ of Claim 2.1.1 naturally corresponds to both the δ_3 and the ω_3 of Claim 1.6.2.) In Chapter 4 we will show that if A is a β_ϕ-connected set of grid points in F_ϕ, then the topological interior of the union of voxels associated with the points in A is topologically connected.

Figure 2.1.5 also illustrates an improvement over a cubic grid from the point of view of the second stated disadvantage of those grids: since two faces of voxels in an fcc grid which meet at an edge are such that the normals to them are 60° from each other, the computer graphic display of surfaces based on the fcc grid appear to be much less blocky than those based on the cubic grid with voxels of the same size. This is further illustrated in Figure 2.1.1.

There may be (and, in fact, should be) a suspicion in the reader's mind as to whether or not the example of Figure 2.1.1 is representative. Maybe if we used a different shape (or even just different angles of tilt and rotation), we would have ended up with a figure illustrating the opposite conclusion. One example immediately comes to mind: the surface of a cube whose faces lie exactly halfway between the grid points of a cubic grid will have a perfect digital approximation in that cubic grid but not in any fcc grid. (For example, consider the cube $C = \{ v \in R^3 \mid -2.5 \le v_i \le 2.5, \text{ for } 1 \le i \le 3 \}$. This cube is exactly the union of the Voronoi neighborhoods of those 125 voxels in the grid Z^3 which are associated with the grid points in the set $\{ c \in Z^3 \mid -2 \le c_i \le 2, \text{ for } 1 \le i \le 3 \}$. There is no way to select grid points in F_ϕ, for any $\phi > 0$, so that the union of the Voronoi neighborhoods in F_ϕ of these grid points will be exactly the cube C.) It turns out that it is this cube example which is less representative of the general situation. To justify a general statement of the superiority of the fcc grids over the cubic grids from the point of view of digitally approximating continuous functions, we would need to diverge into a discussion of sampling theory, which in its turn is based on the theory of Fourier transforms. This would take us too far away from our main line of development and so, rather than getting into a rigorous justification, we restrict ourselves to a vague discussion. A reader who is interested in the mathematically precise version of the contents of the next paragraph may wish to look at Theorem 1.5 on page 62 of [32] (and the material surrounding it).

Assume, as is quite reasonable, that the function f over R^3 which is the object of our interest has a finite support (i.e., it is zero-valued outside a bounded region of R^3). Hence, for any grid, we know that the value of the function is zero at grid points not in the support. Therefore the cost of specifying f at the grid points is proportional to the number of grid points inside the support. Assuming that the support is something reasonable (such as a cube or a ball), it is clear that the costs of specifying f at the points of the cubic grid δZ^3 and the fcc grid $F_{\delta/\sqrt[3]{2}}$ are approximately the same. There is a standard method of estimating the values of f at arbitrary points from its values at such grid points; see, e.g., page 62 of [32]. Also, assuming that f satisfies certain conditions, there is an error bound on the difference at an arbitrary point in R^3 between the value of f and its standard approximation at that point. This error bound is always smaller for the fcc grid $F_{\delta/\sqrt[3]{2}}$ than for the cubic grid δZ^3. Commonly, this is expressed by saying that the fcc grids are more *efficient* than the cubic grids.

Now we discuss a potential advantage of the fcc grids from the point of view of one of the main concerns of this book: boundary tracking. (In Chapter 8 we will see that this potential advantage can indeed be turned into an actual advantage.) As we have seen in the previous chapter (see especially Figure 1.7.2), an algorithm which causes a Fat Fly to visit every face of an arbitrary object, must cause the Fat Fly to traverse every edge at least once and, consequently, to visit every face twice. Since in an fcc grid two voxels which share an edge must in fact share a whole face (see Figure 2.1.5) a situation corresponding to that depicted in Figure 1.7.2 cannot arise, and so it may indeed be possible to track all the faces of an object without visiting each face twice. For example, in Figure 2.1.6 we give two different routes for a Fat Fly for visiting every face of a single rhombic dodecahedron exactly once. In either case, there does not exist a way of attaching an adjacent rhombic dodecahedron to the given one so that the Fat Fly following the arrows would not crawl onto a face of this second rhombic dodecahedron. By itself, unfortunately, this does not solve the boundary tracking problem for fcc grids, since the indicated routes of the Fat Fly for two

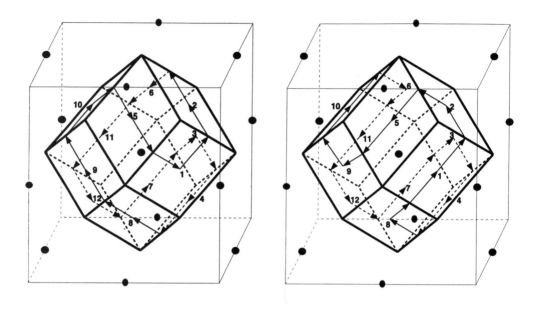

Figure 2.1.6. Two possible routes of a Fat Fly on a single rhombic dodecahedron (compare with Figure 1.7.1).

adjacent rhombic dodecahedra are not consistent with each other: the end of an arrow on a face of one of them does not necessarily correspond to the beginning of an arrow on the face of the other onto which the Fat Fly is supposed to have crawled. We delay the resolution of this difficulty till Chapter 8.

2.2. Other Spaces

In the next chapter we give a formal definition of digital spaces as a pair consisting of a set and an adjacency on the set. Following that definition, we will be in position to define mathematically a variety of digital spaces. In this section we provide more informal (essentially geometrical) descriptions of such spaces. The purpose of this is to motivate the formal definition of a digital space. We wish to illustrate here the great diversity of things that have to be covered by the single definition of a digital space and thereby justify its rather general character.

We begin with the spaces based on the square grid Z^2. We have already seen two different adjacencies defined on this grid, the 4-adjacency ω_2 and the 8-adjacency $\alpha_2 = \delta_2$. Other adjacencies have also been studied in the literature. For example, Preston [37] advocates the use of 6-adjacencies (which arise if we use one but not both diagonals) as the ones most appropriate for image processing using cellular logic. There are two such adjacencies; they can be defined as

$$\beta = \{\, (c,d) \,|\, (c,d) \in \alpha_2 \ \& \ (c_1 - d_1) \neq (d_2 - c_2) \,\} \tag{2.2.1}$$

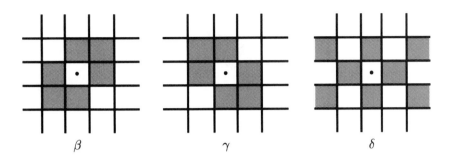

Figure 2.2.1. The adjacencies β, γ, and δ. The shaded pixels form exactly the set of all pixels adjacent to the pixel labeled •.

and

$$\gamma = \{ (c,d) \mid (c,d) \in \alpha_2 \ \& \ (c_1 - d_1) \neq (c_2 - d_2) \} . \tag{2.2.2}$$

We will also have occasion to use an adjacency which allows certain "knight moves" and is defined as

$$\delta = \{ (c,d) \mid [\, |c_2 - d_2| = 1 \ \& \ (c_1 = d_1 \text{ or } |c_1 - d_1| = 2) \,] \\ \text{or } [\, |c_1 - d_1| = 1 \ \& \ c_2 = d_2 \,] \} . \tag{2.2.3}$$

In Figure 2.2.1 we indicate the pixels which are β-, γ-, and δ-adjacent to a given pixel.

Staying in the plane R^2, we may consider grids which are not square. An obvious possibility is to consider the formal equivalent of the fcc grids, namely, for a positive ϕ

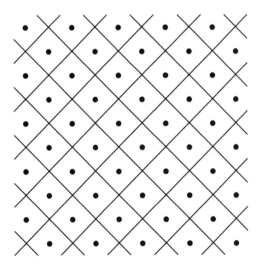

Figure 2.2.2. The grid G_ϕ and the associated Voronoi neighborhoods. Such a configuration can also be obtained by rotating the square grid $\sqrt{2}\phi Z^2$ by $45°$ (compare Figure 1.1.2).

define

$$G_\phi = \{(\phi c_1, \phi c_2) \mid c_1, c_2 \in Z \text{ and } c_1 + c_2 \equiv 0 \pmod{2}\}. \tag{2.2.4}$$

This does not lead, however, to something essentially new. In Figure 2.2.2 we show this grid and the associated Voronoi neighborhoods; it is clear that these are the same as those provided by the cubic grid $\sqrt{2}\phi Z^2$, after a rotation of $45°$.

In fact, it turns out that, in some other senses, the appropriate two-dimensional analog of the fcc grids is not G_ϕ, but rather the so-called *hexagonal grids*. (These are grids for which the resulting Voronoi neighborhoods are all translates of a particular regular hexagon.) One reason for this claim, which we cannot detail here for lack of the necessary mathematical background, has to do with the efficiency of the hexagonal grids. We simply state here that the hexagonal grids are more efficient than the square grids, in the same sense as the fcc grids are more efficient than the cubic grids (see, e.g., [36]). The discussion of another sense in which the hexagonal grids correspond to the fcc grids is postponed till Chapter 3, where we discuss the notion of adjacencies defined using the so-called N-dimensional sign functions.

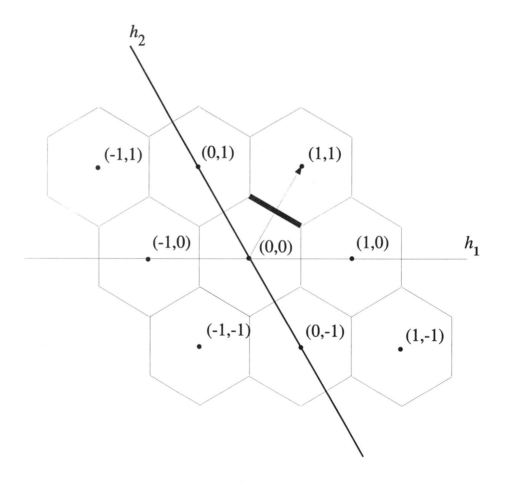

Figure 2.2.3. The hexagonal grid H together with the Voronoi neighborhoods of the grid points.

All hexagonal grids can be obtained by translation, rotation, and scaling from the grid

$$H = \left\{ h_1 \times (1,0) + h_2 \times (-0.5, \sqrt{0.75}) \mid (h_1, h_2) \in Z^2 \right\}. \qquad (2.2.5)$$

This grid is shown in Figure 2.2.3. In the figure we show some of the grid points and the corresponding Voronoi neighborhoods (pixels). We also indicate two other ideas. One is that in addition to the coordinates in R^2, the grid points can also be represented by the pairs (h_1, h_2) of integers used in the definition of H above. Thus the grid point at $(0.5, \sqrt{0.75})$ is labeled in the figure as $(1,1)$. Now if we consider the adjacency in the hexagonal grid defined by the corresponding Voronoi neighborhoods having exactly one edge in common (edge-adjacency, to be denoted by ϵ), then we see the potential for identifying the hexagonal grid H with the square grid Z^2 and edge-adjacency ϵ with the adjacency β; compare the left side of Figure 2.2.1 with Figure 2.2.3. This will provide one of the examples of the isomorphisms to be discussed in Chapter 3. The second idea emphasized in Figure 2.2.3 is the intuitive geometrical interpretation of an adjacency. The adjacent pair of grid points $\big((0,0), (0.5, \sqrt{0.75})\big)$ is not to be interpreted as the arrow pointing from the first to the second, but rather as the edge (shown dark) shared by the Voronoi neighborhoods of the two grid points, with an orientation across it from the first grid point to the second; compare the discussion in the second paragraph after Claim 1.5.1.

Figure 2.2.3 also makes obvious a third similarity between the fcc grids and the hexagonal grids: just as in the fcc grids there is no face-or-edge-adjacency which is not also a face-adjacency (see Figure 2.1.5), in the hexagonal grid there is no edge-or-vertex-adjacency, which is not also an edge-adjacency. If two pixels in the hexagonal grid share a vertex, they must also share an edge. It will also be shown to be the case that the result for hexagonal grids corresponding to Claim 1.5.2 for square grids is true when both α_2 and ω_2 are replaced by edge-adjacency for the hexagonal grid. It is of course possible to define other adjacencies for the hexagonal grids. For example, to demonstrate some further desirable properties of them as opposed to the square grids, we will have (in Chapter 5) an exercise which uses an adjacency involving knight moves on a hexagonal grid, similar to the δ illustrated on the right side of Figure 2.2.1.

Clearly, there are other ways of tessellating the plane into polygons of various kinds and for every one of these it is possible to define notions of edge-adjacency and edge-or-vertex-adjacency; see Exercise 1.5. We do not pursue this thought further here, but move onto R^3.

Although we will need to introduce adjacencies in addition to the already discussed α_3, δ_3, and ω_3 for the cubic grids (see Figure 1.6.2), such as the three-dimensional versions of β and γ of Figure 2.2.1, we will postpone doing this until the need arises. In the case of the fcc-grids, in addition to β_ϕ which is both the face-adjacency and the face-or-edge-adjacency for such grids, one may consider an additional face-or-edge-or-vertex-adjacency. Looking at Figure 2.1.2, we see that the voxel of the grid point g has exactly one vertex in common with the voxels of six other grid points, meeting them at the centers of the faces of the $2 \times 2 \times 2$ cube shown in the figure. Intuitively, the connectivity of two voxels which are adjacent in this fashion is very flimsy indeed. Nevertheless, we will find such an adjacency useful in our discussion of boundary tracking in the fcc grid (Chapter 8).

Clearly, there are other ways of tessellating the three-dimensional space into polyhedra of various kinds (a *polyhedron* is a solid bounded by plane polygons, which are referred to as its *faces*), and for every one of these it is possible to define notions of face-adjacency, face-or-edge-adjacency, and face-or-edge-or-vertex-adjacency. We mention only one of these in detail because it arises from grids which are well known, together with the cubic grids and the fcc grids, in many areas of science, such as geometry [7], solid state physics [20], sampling theory [36], and image reconstruction from projections [29]. These are the *body-centered cubic grids* (or *bcc grids*, for short), defined for any positive real number ϕ by

$$B_\phi = \{(\phi c_1, \phi c_2, \phi c_3) \mid c_1, c_2, c_3 \in Z \text{ and } c_1 \equiv c_2 \equiv c_3 \pmod 2\}. \qquad (2.2.6)$$

These grids are more efficient than the fcc grids (in the same sense as the fcc grids are more efficient than the cubic grids, see, e.g., [36]), and they have been advocated as the most appropriate grids for the three-dimensional euclidean space [27]. Nevertheless, it so happens that we will not be using these grids in any of the illustrations of our general theory, and so we restrict ourselves here to briefly describing some of their properties.

The reader should find it easy to verify that the voxels associated with these bcc grids are truncated octahedrons having eight hexagonal and six square faces and, hence, for every grid point there are 14 grid points face-adjacent to it. It should also be not too difficult to show that no additional grid points are adjacent to a given one in the face-or-edge or in the face-or-edge-or-vertex sense, since whenever two voxels share a vertex (or an edge), they must also share a whole face. It requires more work to show why it is that the bcc grids are alternative (to the fcc grids) three-dimensional generalizations of the hexagonal grids. We will do this in the next chapter. It is also the case that the bcc and fcc grids are "reciprocal grids" in a mathematically well-defined sense [7, 20], but a detailed discussion of this would take us too far from the topic of this book.

In what follows, we will also have occasion to use various higher dimensional generalizations of some the digital spaces described so far. Such generalizations can have practical motivations. For example, the fourth dimension in R^4 can be interpreted as time, and one can have digital objects in R^4 which are interpreted as sampled (in space and time) versions of time-varying continuous objects, such as the human heart or lungs [17]. We will introduce such digital spaces as we need them. Here we complete this section with a discussion of some digital spaces which do not originate from a set of grid points in R^N.

The examples we give are based on tessellations of space into cubes (i.e., on cuberilles). It will be easily seen, however, that the ideas introduced can be applied to other tessellations as well. Consider the situation presented in Section 1.2: a finite number of voxels are occupied by sugar cubes. Consider further, for example, a δ_3-component of the occupied set of voxels. Let us refer to this, following the terminology of Section 1.2, as an "object." Continuing with the terminology of Section 1.2, let us call a face which separates a voxel in the object from a voxel not in the object a "boundary face." Each boundary face has four edges and at each edge one of three possibilities may arise (these are illustrated in Figure 2.2.4, in which the given boundary face is shaded darkly and the edge in question is drawn heavily):

(i) The edge in question is not in any other voxel of the object.

(ii) The edge in question is in one other voxel of the object and that voxel shares a face with the voxel of the given boundary face.

(iii) The edge in question is in another voxel of the object and that voxel shares only that edge with the voxel of the given boundary face.

As depicted in Figure 2.2.4, in each of the three cases there is a unique boundary face (shaded lightly) to which the given darkly shaded face is considered adjacent. We leave it to the reader to show that this binary relation is indeed symmetric (i.e., that it is also the case that the lightly shaded boundary face is adjacent to the darkly shaded one).

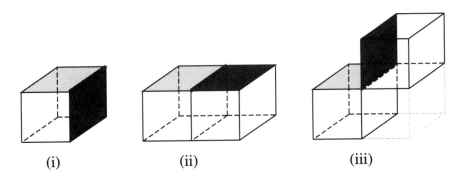

(i)   (ii)   (iii)

Figure 2.2.4. The darkly shaded boundary face of an object is "adjacent to" another boundary face at each of its edges. The location of this adjacent face, shaded lightly, depends on which other voxels in the object contain the edge in question. (In case (iii), it does not matter whether or not the voxel drawn very lightly is occupied by a sugar cube.)

Now consider a component in the set of all the boundary faces of an object under the adjacency defined in the manner indicated by Figure 2.2.4. Such a component with such an adjacency will be an example of a digital space as formally defined in the next chapter. It will also be "the boundary containing any of the boundary faces in it," a concept whose well definedness we have considered open to question in Section 1.2. It looks like we are making progress! In fact, as we will see, the boundary $\partial(O, Q)$ in Aim 1.6.1 is exactly the "boundary containing the boundary face (g, h)."

Now we have considered digital spaces whose elements are voxels and in which adjacent voxels share a face and also digital spaces whose elements are voxel faces and in which adjacent faces share an edge. We can combine these notions and consider a digital space, based on a cuberille, in which the elements are either voxels or faces of voxels or edges of

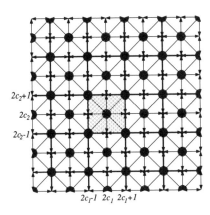

Figure 2.2.5. A pixel in $2Z^2$ is indicated by shading. The vertices and the centers of edges of such a pixel correspond to grid points of Z^2; compare with Figure 1.1.2. The Khalimsky adjacencies are indicated by the lines with arrows. (Eight arrows meeting at a point give the appearance of a circle.)

voxels or vertices of voxels. We may define adjacency as follows: two elements are adjacent if, and only if, one is a proper subset of the other (e.g., a voxel is adjacent to all its vertices). It turns out that this space is isomorphic (in the sense to be made precise in Chapter 3) to a space whose elements are Z^3 with a special kind of adjacency. For ease of illustration, we discuss here only the two-dimensional version of this space.

In the two-dimensional case we start with the square grid $2Z^2$. The vertices of a pixel and the centers of edges of a pixel of the grid $2Z^2$ are grid points of Z^2 (see Figure 2.2.5), and the set K consisting of pixels, edges of pixels, and vertices of pixels is clearly in a one-to-one correspondence with Z^2. The adjacency in Z^2 which corresponds to the symmetric binary relation on K described by "one of the elements is a proper subset of the other" is referred to as a *Khalimsky adjacency*. It is represented in Figure 2.2.5 by arrows between the corresponding points of Z^2. The resulting digital space has some mathematically desirable properties which will be further discussed below, especially in Chapters 4 and 7.

2.3. Exercises

2.1.   Prove that both the fcc grid and the bcc grid are Delone grids in R^3.

2.2.   Consider a point v in R^3 such that there are two voxels in the fcc grid which share v but do not share a face. How many other voxels also contain v?

2.3.   Sketch the appearance of the Voronoi neighborhoods in the bcc grid, and show that two voxels which share a vertex must also share a whole face.

2.4.   Prove that two grid points of the bcc grid B_1 are adjacent if, and only if, either they differ only in one of their components and the absolute value of the difference in that component is 2 or they differ in all three components and the absolute values of the differences in each of the components is 1.

2.5.   Prove that the binary relation defined on the set of boundary faces by the rules depicted in Figure 2.2.4 is symmetric.

2.6.   Define precisely the Khalimsky adjacency for Z^3.

3
Digital Spaces

"Man muß nicht alles für wahr halten, man muß es nur für notwendig halten."
("One need not consider it all to be true, one must only consider it to be necessary.")

F. Kafka, *Der Prozess (The Trial)*, Ninth Chapter.

3.1. The Basic Definitions

Our aim is to provide a mathematical framework for a theory of objects and their surfaces which is applicable to multidimensional discrete spaces. Our motivation comes from practical applications in which boundaries need to be identified in multidimensional data sets with the further aim of displaying them on a computer screen (see Figure 1.1.3). Our definitions are biased towards such applications. One of our aims is to characterize surfaces with a well-determined inside and outside and to define boundaries of objects so that they are indeed surfaces of this type. (In particular, this means that our surfaces must have an orientation, so that we can tell which side is the inside as opposed to the outside.) Furthermore, we want to make our presentation general enough to incorporate many of the reasonable but ad hoc ways that notions, such as "connectedness" and "boundary," may be defined in digital geometry. We need a framework appropriate for a mathematical treatment of the intuitive notion of a "surface with a connected inside and a connected outside" (a "Jordan surface") in the discrete multidimensional environment.

We do not wish our theory to be restricted to the type of approach based on the N-dimensional euclidean space R^N which is assumed to be digitized according to a tessellation by N sets of parallel planes. Each element in such a digitization can be represented by an N-dimensional vector of integers, an element of Z^N (similarly to a cubic grid). As seen in the previous chapter, there are other potentially more useful approaches which would not be covered by a framework based solely on such tessellations.

The current chapter is the beginning of our formal introduction to this general theory. It will turn out that the notion of a "Jordan surface" (and the theorems associated with that notion) can be incorporated into an extremely general framework, one which does not even assume that there is an underlying euclidean space. Thus the results apply to a wide variety of geometrically meaningful special cases (and, accidentally, to some which make no sense at all from the geometrical point of view). The important claims made earlier become corollaries to the results proven below.

Now we define the basic concept of our book. A *digital space* is a pair (V, π), where V is an arbitrary nonempty set and π is a symmetric binary relation on V such that V is π-connected. (The concepts used in this definition have been defined in Section 1.4.) Sometimes we will refer to π as the *proto-adjacency* of the digital space (V, π).

Note that the definition of a digital space is much more general than is necessary to do "digital geometry," since there is no geometrical restriction on the nature of the set V. (V, π) is essentially just a connected digraph, except that we allow V to be infinite and a node to be adjacent to itself. Even (V, V^2) is a legitimate example. (V^2 denotes the set of all ordered pairs of elements of V, i.e., it is the binary relation that holds between any two elements of V.) Although this will not enter the mathematical discussion at all, the way we usually imagine these spaces is much more restrictive. We think of an N-dimensional euclidean space R^N and some arbitrary but fixed grid in this space. Then we think of V as the set of the Voronoi neighborhoods of the grid points, as defined in (1.1.3), and allow two elements of such a set V to be in the relation π only if their intersection is $(N - 1)$-dimensional. In the three-dimensional case, this means that two Voronoi neighborhoods (which in this case can also be referred to as voxels) can be in π only if they share a face.

Although the geometrical idea described above will have no influence on how our proofs are organized, it does influence the terminology we have decided to adopt: elements of V will be called *spels* (short for spatial elements) and elements of π will be called *surfels* (short for surface elements). Any nonempty subset of π will be called a *surface* in (V, π).

Note that if (c, d) is in a surface, (d, c) need not be. This is an essential feature of the theory that we are trying to develop. This theory is supposed to deal with surfels oriented so that we can distinguish between their "insides" and their "outsides." Most of our results apply to the rather general class of surfaces as defined in the previous paragraph. Very occasionally we will need to restrict our attention to *antisymmetric surfaces* S, meaning that if $(c, d) \in S$, then $(d, c) \notin S$. (Note that an antisymmetric surface is automatically *antireflexive*: for any spel c, $(c, c) \notin S$.) Even though we will not need antisymmetry for most of our proofs, our examples of surfaces will usually be antisymmetric. (Indeed, surfaces which naturally occur in an application of our theory are most likely to be antisymmetric.)

Let us dwell on our chosen terminology a little longer. Since a digital space is essentially a (possibly infinite) digraph (with a symmetric and possibly reflexive adjacency relation), the question arises why we do not use a purely graph-theoretical terminology. For example, instead of "spel" we could use "node." The reason for not doing that is our geometrical orientation. Many of the digital spaces of practical importance are defined using Voronoi neighborhoods by letting V be a grid in R^N ($N \geq 1$) and π be the set of ordered pairs of distinct points in V such that the nonempty intersection of their Voronoi neighborhoods is not also a subset of the Voronoi neighborhood of a third grid point. (When $N = 2$, this means that the Voronoi neighborhoods share at least a line segment, i.e., they are two polygons which have exactly a single edge in common; see Exercise 1.5. When $N = 3$, the implication is that the Voronoi neighborhoods are two polyhedra with exactly a single face in common. These are, respectively, the already alluded to notions of edge-adjacency and face-adjacency.) The term spel, short for "spatial element," reflects well this geometrical interpretation in terms of Voronoi neighborhoods. (It corresponds to the terms pixel for R^2 and voxel for R^3.) Further, the question also arises as to why we refer to the "arcs" of the graph which is the digital space as "surfels." Again the answer lies in the geometrical interpretation. In Figure 2.2.3, the "arc"

$((0,0),(1,1))$ is indicated by the arrow from $(0,0)$ to $(1,1)$. However, this is not how we think of $((0,0),(1,1))$. In our mind $((0,0),(1,1))$ is interpreted as the heavily drawn edge of the hexagonal Voronoi neighborhood of $(0,0)$ which is shared with the Voronoi neighborhood of $(1,1)$ (with an orientation from $(0,0)$ to $(1,1)$). Similarly, it is better not to interpret (g,h) in Figure 2.1.5 as a vector from the grid point g to the grid point h. In our minds (and in the programs which produce images such as shown in Figure 1.1.3), (g,h) is interpreted as that face of the rhombic dodecahedral Voronoi neighborhood of g shared with the Voronoi neighborhood of h (with an orientation from g to h). Hence it is appropriate to talk about surfels, short for "surface elements"; the word arc is not nearly so descriptive. The orientation of the surfel is important when using its description. For example, in producing computer graphic displays, such as in Figure 1.1.3, the known orientation of the surfels reduces our labor by at least 50%. The surfels oriented away from the viewer need not be considered at all, since they are going to be hidden. (This is a consequence of the Jordan properties which we discuss below.) Also, volumes within antisymmetric surfaces can be obtained by using the orientation and a digital version of Gauss's theorem [19].

We will be particularly concerned with a special class of surfaces, which we refer to as *boundaries*. These are defined as follows. Let (V,π) be a digital space, and let O and Q be subsets of V. Then the *boundary in* (V,π) *between* O *and* Q is defined as

$$\partial(O,Q) = \{ (c,d) \mid (c,d) \in \pi, c \in O \ \& \ d \in Q \} . \tag{3.1.1}$$

(Note that this is a generalization of the definition provided in (1.5.1), which is the special case of the definition above for the digital space (Z^N, ω_N). This generalization provides the promised justification for the use of $\partial(O,Q)$ in Claim 2.1.1. Since the definition of the boundary depends on the proto-adjacency of the digital space, it would have been reasonable to use the notation ∂_π instead of just ∂. However, in our work there is typically only one proto-adjacency that we deal with at a time. Therefore we do not have to complicate our notation for a boundary with an additional suffix. We will introduce the more cumbersome notation when we are forced to do so.) If it is not empty, then a boundary in (V,π) is a surface in (V,π). In what follows, as is usual in the literature, we will normally be dealing with nonempty boundaries between disjoint sets. Clearly, such boundaries are antisymmetric surfaces.

For simplicity in illustrating our ideas, in many of our figures we use $V = Z^2$, a square grid. The intuitive geometrical interpretation of this is that the underlying space is the plane R^2 and the grid points are the points with integral coordinates. Thus the corresponding Voronoi neighborhoods (pixels in this case) are the closed unit squares whose centers are at the grid points. An ordered pair of pixels which share an edge is interpreted as that edge with an orientation across it from the first pixel to the second one. The corresponding digital space is (Z^2, ω_2), much discussed in Chapter 1 and illustrated in Figure 1.1.2.

An in-practice much less useful but (for demonstrating the generality of our concepts) more interesting example is (Z^2, χ), where

$$\chi = \{ ((c_1,c_2),(d_1,d_2)) \mid [\ |c_2 - d_2| = 1 \ \& \ c_1 = d_1 = 0\ ]$$
$$\text{or}\ [\ |c_1 - d_1| = 1 \ \& \ c_2 = d_2\ ] \} . \tag{3.1.2}$$

A pixel is χ-adjacent to its vertical neighbors only if it lies on the vertical axis; otherwise a pixel is χ-adjacent only to its two horizontal neighbors (see Figure 1.1.2). Nevertheless,

(Z^2, χ) is a digital space, since χ is symmetric and Z^2 is χ-connected. (We can go from any pixel to any other pixel by first moving horizontally till the vertical axis is reached, then moving along the vertical axis until we get to the horizontal location of the pixel to be reached, and then finally reaching that pixel by horizontal steps.)

3.2. Interiors and Exteriors

Let (V, π) be a digital space, and let S be a surface in it. We define the *immediate interior* $II(S)$, the *immediate exterior* $IE(S)$, and the *immediate neighborhood* $IN(S)$ of S as follows:

$$
\begin{aligned}
II(S) &= \{\, c \mid (c,d) \in S \text{ for some } d \text{ in } V \,\}\,, \\
IE(S) &= \{\, d \mid (c,d) \in S \text{ for some } c \text{ in } V \,\}\,, \\
IN(S) &= II(S) \cup IE(S)\,.
\end{aligned}
\tag{3.2.1}
$$

We say that a π-path $\langle c^{(0)}, \cdots, c^{(K)} \rangle$ *crosses* the surface S if there is a k, $1 \le k \le K$, such that either $(c^{(k-1)}, c^{(k)}) \in S$ or $(c^{(k)}, c^{(k-1)}) \in S$. The *interior* $I(S)$ and the *exterior* $E(S)$ of S are defined as follows:

$$
\begin{aligned}
I(S) &= \big\{\, c \in V \mid \text{there exists a } \pi\text{-path connecting } c \text{ to an} \\
&\qquad\qquad \text{element of } II(S) \text{ which does not cross } S \,\big\}\,, \\
E(S) &= \big\{\, c \in V \mid \text{there exists a } \pi\text{-path connecting } c \text{ to an} \\
&\qquad\qquad \text{element of } IE(S) \text{ which does not cross } S \,\big\}\,.
\end{aligned}
\tag{3.2.2}
$$

Lemma 3.2.1. *For any surface S in a digital space (V, π),*

$$
I(S) \cup E(S) = V\,.
\tag{3.2.3}
$$

Proof. Let c be an arbitrary element of V and d be an arbitrary element of $IN(S)$. (By definition, neither of the sets V and $IN(S)$ is empty.) Since V is π-connected, there exists a π-path $\langle c^{(0)}, \cdots, c^{(K)} \rangle$ from c to d. Let k be the smallest nonnegative integer such that $c^{(k)}$ is in $IN(S)$. Then $\langle c^{(0)}, \cdots, c^{(k)} \rangle$ is a π-path connecting c to an element of $II(S) \cup IE(S)$ which does not cross S. Hence c is in at least one of $I(S)$ and $E(S)$. $\square$

In this proof we have made essential use of the π-connectedness of V. It is easy to see that the lemma would not be true if we did not insist that V be π-connected.

A surface S in a digital space (V, π) is said to be *near-Jordan* if every π-path from any element of $II(S)$ to any element of $IE(S)$ crosses S. (This immediately implies that a near-Jordan surface is an antisymmetric surface.) In Figures 3.2.1 and 3.2.2 we illustrate this most important concept for the space (Z^2, χ). It is easy to see (observe Figure 3.2.1) that for this digital space any surface which consists of exactly one element is near-Jordan. We also see that the interior and the exterior of the near-Jordan surface of Figure 3.2.1 partition

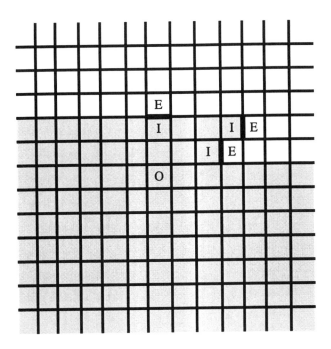

Figure 3.2.1. Example of a near-Jordan surface for the digital space (Z^2, χ) with the spel at the origin labeled O. The surface consists of only three surfels represented by bold edges. The three spels in the immediate interior are labeled I, and the three spels in the immediate exterior are labeled E. The spels in the interior are shaded, and the spels in the exterior are not. This illustrates Lemma 3.2.2, namely that the interior and the exterior of a near-Jordan surface partition the set of spels. (Reproduced from [16] with the publisher's permission.)

the set of all spels. That this is true for all near-Jordan surfaces is shown in the next lemma. We remark here that if S is not a near-Jordan surface, then the intersection of its interior and its exterior is necessarily nonempty. (This is because there is a path, not crossing S, from its immediate interior to its immediate exterior. All the spels in this path are in both the interior and the exterior of S, as can be seen from (3.2.2), the symmetry of proto-adjacency, and the symmetrical definition of "crosses.") One might be tempted to conjecture that in this case both the interior and the exterior must be the whole space V. This is not so, however. A counterexample is given in Figure 3.2.2.

Lemma 3.2.2. *Let S be a surface in a digital space (V, π). Then the following three conditions are equivalent.*

   (i) *S is near-Jordan.*

   (ii) *Every π-path from any element of $I(S)$ to any element of $E(S)$ crosses S.*

   (iii) *$I(S) \cap E(S) = \emptyset$. (As usual, $\emptyset$ denotes the empty set.)*

Furthermore, if these conditions are satisfied, then it is also the case that

$$S = \partial(I(S), E(S)) \,. \tag{3.2.4}$$

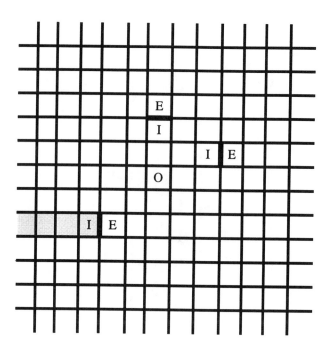

Figure 3.2.2. Example of a surface in the digital space (Z^2, χ) that is not near-Jordan, with the spel at the origin labeled O. The surface consists of only three surfels represented by bold edges. The three spels in the immediate interior are labeled I, and the three spels in the immediate exterior are labeled E. From either of the upper two spels in the immediate interior there is a χ-path to the lowest spel in the immediate exterior which does not cross the surface. There is a nonempty intersection between the interior and the exterior but neither contains all the spels. The spels in the shaded region are not in the exterior, and the spels in the upper part of the space are not in the interior. (Reproduced from [16] with the publisher's permission.)

Proof. Let S be any surface in a digital space (V, π). Suppose that there is a π-path $\langle c^{(0)}, \cdots, c^{(K)} \rangle$ not crossing S connecting a c in $I(S)$ to a d in $E(S)$. By (3.2.2), there is also a π-path $\langle e^{(0)}, \cdots, e^{(L)} \rangle$ not crossing S connecting c to an element of $II(S)$ and a π-path $\langle d^{(0)}, \cdots, d^{(M)} \rangle$ not crossing S connecting d to an element of $IE(S)$. Then $\langle e^{(L)}, \cdots, e^{(0)} = c = c^{(0)}, \cdots, c^{(K)} = d = d^{(0)}, \cdots, d^{(M)} \rangle$ is a π-path not crossing S which connects an element of $II(S)$ to an element of $IE(S)$. Therefore, by definition, S is not near-Jordan. This argument shows that (i) implies (ii). If $I(S)$ and $E(S)$ have an element c in common, then the trivial π-path $\langle c \rangle$ is from an element of $I(S)$ to an element of $E(S)$ and does not cross S. This shows that (ii) implies (iii). That (iii) implies (i) was essentially proven in the paragraph preceding the statement of the lemma. Finally, it is the case for any surface S that it is a subset of $\partial(I(S), E(S))$. (This follows trivially from the definitions of immediate interior and exterior and of interior and exterior.) Now suppose that (i)-(iii) hold and that the surfel (c, d) is in $\partial(I(S), E(S))$. Then d belongs to $E(S)$, and so by (iii) d does not belong to $I(S)$. Hence (d, c) is not in S. The π-path $\langle c, d \rangle$ is from an element of $I(S)$ to an element of $E(S)$, and so, as condition (ii) is satisfied, it crosses S. Hence (c, d) is in S. This proves that any of the conditions (i)-(iii) implies (3.2.4). $\square$

We note that (3.2.4) does not imply conditions (i)-(iii), as can be easily seen by considering the case when the surface S is chosen to be the whole of π.

Lemmas 3.2.1 and 3.2.2 say that the interior and the exterior of a near-Jordan surface partition V. Furthermore, the surface itself lies on the boundary of the interior and the exterior, and we cannot get from the interior to the exterior without crossing through the surface. All this sounds like a "Jordan Curve Theorem" except for one essential point: nothing is said about the interior and the exterior being connected sets. This is not surprising, because although the simple closed curves of the Jordan Curve Theorem are by definition "connected" (recall that they were supposed to have been drawn "continuously"), there is nothing in our definitions which implies that a near-Jordan surface is in any sense "connected." In the next section we consider connectedness of the interior and the exterior in our framework.

There is one other aspect of the Jordan Curve Theorem which is missing from Lemmas 3.2.1 and 3.2.2. It has to do with the fact that a simple closed curve is a subset of the plane and the Jordan Curve Theorem says that the complement of the curve, rather than the whole plane, has precisely two components. Since our surfaces are subsets of π, rather than of V, it is possible to demand that the interior and exterior partition the whole of V; if anything, a more attractive-sounding aim than that of the classical theorem.

Another property of simple closed curves in the plane is that exactly one component of the complement of such a curve is bounded. For Delone grids, such as Z^N (see Exercise 1.2), the continuous notion of boundedness corresponds to the digital notion of finiteness (see Exercise 1.4). It is a consequence of the material in the following paragraphs that, for $N \geq 2$, if S is a finite near-Jordan surface in $\left(Z^N, \omega_N\right)$, then exactly one of $I(S)$ or $E(S)$ is finite. However, this result does not carry over to the more general framework of this book, even if we restrict V to be Z^2. An example is provided in Figure 3.2.1, in which a near-Jordan surface has only three elements, but both its interior and its exterior are infinite.

To state our promised finiteness results, we need a technical definition. We call (V, π) a *boundable digital space* if V is a Delone grid in R^N for some positive N and, for every nonnegative integer C, the set $\{\, c \in V \mid \|c\| \geq C \,\}$ is π-connected.

Theorem 3.2.3. *For $N \geq 2$, $\left(Z^N, \omega_N\right)$ is a boundable digital space.*

Proof. We left it to the reader to prove (in Exercise 1.2) that Z^N is a Delone grid in R^N. We need to prove that the set $O = \{\, c \in Z^N \mid \|c\| \geq C \,\}$ is ω_N-connected. To do this, we define a function sgn on the real numbers by

$$\mathrm{sgn}(r) = \begin{array}{ll} +1, & \text{if } r \geq 0, \\ -1, & \text{if } r < 0. \end{array} \tag{3.2.5}$$

To prove the ω_N-connectedness of O, it is clearly sufficient to prove the following two statements:

(i) Any $c \in O$ is ω_N-connected in O to $(\mathrm{sgn}(c_1)C, \cdots, \mathrm{sgn}(c_N)C)$.

(ii) Any $c \in Z^N$ such that, for $1 \leq n \leq N$, either $c_n = C$ or $c_n = -C$ is ω_N-connected in O to $(C, \cdots, C)$.

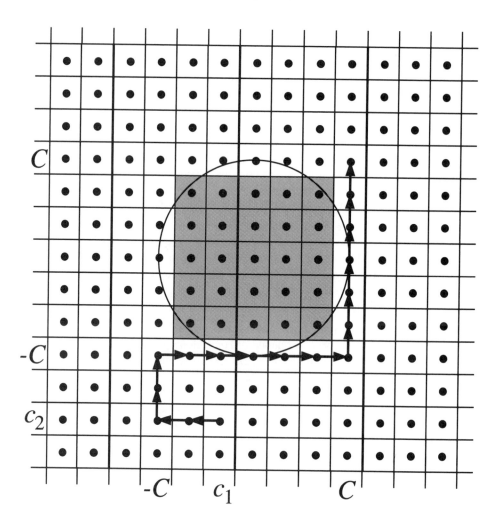

Figure 3.2.3. An illustration of the ideas in the proof of Theorem 3.2.3. Here $N = 2$, $C = 3$, and the Voronoi neighborhoods of all those elements of Z^2 which are not in O are shaded. Further, $c_1 = -1$ and $c_2 = -5$. The figure indicates an ω_2-path in O, from c to $(-C, c_2)$, then to $(-C, -C)$, then to $(C, -C)$, and finally to (C, C).

The ideas involved in proving these statements are illustrated in Figure 3.2.3

To prove (i), we show that any $c = (c_1, c_2, \cdots, c_N) \in O$ is ω_N-connected in O to $(\mathrm{sgn}(c_1)C, c_2, \cdots, c_N)$. Then the rest of the proof can be carried out in the same way from the second to the Nth coordinate. First consider the case that $|c_1| \leq C$. Then the ω_N-path $\langle c^{(0)}, \cdots, c^{(K)} \rangle$ from c to $(\mathrm{sgn}(c_1)C, c_2, \cdots, c_N)$ in which, for $0 \leq k \leq K$, $c_1^{(k)} = \mathrm{sgn}(c_1)(|c_1| + k)$ and $c_n^{(k)} = c_n$ (for $2 \leq n \leq N$) is in O, since (for $0 \leq k \leq K$) $\|c^{(k)}\| \geq \|c\| \geq C$. The alternative case is that $|c_1| > C$. Then the ω_N-path $\langle c^{(0)}, \cdots, c^{(K)} \rangle$ from c to $(\mathrm{sgn}(c_1)C, c_2, \cdots, c_N)$ in which, for $0 \leq k \leq K$, $c_1^{(k)} = \mathrm{sgn}(c_1)(|c_1| - k)$ and $c_n^{(k)} = c_n$ (for $2 \leq n \leq N$) is in O, since (for $0 \leq k \leq K$) $|c_1^{(k)}| \geq C$ and, hence, $\|c^{(k)}\| \geq C$.

To prove (ii), we also proceed one coordinate at a time; we give the details only for the first coordinate. If $c_1 = C$, then nothing needs to be done for the first coordinate. The alternative is that $c_1 = -C$. Then the ω_N-path $\langle c^{(0)}, \cdots, c^{(2C)} \rangle$ from c to $(C, c_2, \cdots, c_N)$ in which, for $0 \le k \le 2C$, $c_1^{(k)} = (c_1 + k)$ and $c_n^{(k)} = c_n$ (for $2 \le n \le N$) is in O, since (for $0 \le k \le K$) $|c_2^{(k)}| = |c_2| = C$ and, hence, $\|c^{(k)}\| \ge C$. (It is in this last sentence only that we have used the assumption that $N \ge 2$. We leave it to the reader to show that this assumption is necessary.) $\square$

Theorem 3.2.4. *If S is a finite near-Jordan surface in a boundable digital space, then exactly one of $I(S)$ or $E(S)$ is finite.*

Proof. Let S be a finite near-Jordan surface in a boundable digital space (V, π). Then $IN(S)$ is nonempty but has only finitely many elements; see (3.2.1). Let C be an integer greater than $\max\{\|c\| \mid c \in IN(S)\}$. Consider the set $O = \{c \in V \mid \|c\| \ge C\}$ of spels. We have left it to the reader to prove (see Exercise 1.3) that O is infinite. By the definition of a boundable digital space, O is also π-connected. The way C and O have been chosen ensures that a π-path in O does not cross S. Then it follows from (3.2.2) and Lemmas 3.2.1 and 3.2.2 that either $O \subset I(S)$ and $E(S) \subset \overline{O}$ or $O \subset E(S)$ and $I(S) \subset \overline{O}$. To complete the proof, we follow only the first of these two alternatives. The proof for the second alternative is strictly analogous.

Since O is infinite, $I(S)$ is infinite. To complete the proof, we need to show that $E(S)$ is finite, which would certainly follow if we could show that $\overline{O} = \{c \in V \mid \|c\| < C\}$ is finite. We do this by noting that (being a Delone grid) V is a (D, d) grid for some positive real numbers D and d. From this it follows (see Exercise 1.1) that, for any c in V, $N_V(c) \subset B_{D,c}$. This, together with the triangle inequality (which says that, for any v in w in R^N, $\|v + w\| \le \|v\| + \|w\|$) implies that if $v \in \bigcup_{c \in \overline{O}} N_V(c)$, then $\|v\| \le C + D$, i.e., that $\bigcup_{c \in \overline{O}} N_V(c)$ is bounded. This implies that $\overline{O}$ is finite (Exercise 1.4). $\square$

We complete this section with two other characterizations of a near-Jordan surface. The first of these, due to T. Y. Kong, has a significant property that characterizations (ii) and (iii) in Lemma 3.2.2 do not have: it gives an easy way to construct near-Jordan surfaces. It should also help the reader to grasp the essential nature of such surfaces.

Theorem 3.2.5. *Let S be a surface in a digital space (V, π). Then the following three conditions are equivalent.*

(i) *S is near-Jordan.*

(ii) *There exists a nonempty proper subset O of V such that*

$$S = \partial(O, \overline{O}).\tag{3.2.6}$$

(iii) *There exists a nonempty proper subset O of V such that $I(S) = O$ and $E(S) = \overline{O}$.*

Furthermore, if these conditions are satisfied, then the O of (ii) has to be $I(S)$.

Proof. Let S be any surface in a digital space (V, π). Since, by definition, S is nonempty, both $I(S)$ and $E(S)$ are nonempty. If S is also near-Jordan, then it follows from Lemma 3.2.2(iii) that $I(S)$ is a nonempty proper subset of V and $E(S)$ is the complement of $I(S)$ in V. By setting $O = I(S)$, (i) implies (ii) from (3.2.4). Now assume that there is an O that satisfies (ii). To show that for this O we have that $I(S)=O$ and $E(S)=\overline{O}$, it suffices (by Lemma 3.2.1) to show that (3.2.6) implies

$$E(S) \cap O = \emptyset \tag{3.2.7}$$

and

$$I(S) \cap \overline{O} = \emptyset. \tag{3.2.8}$$

First we note that (3.2.6) implies that $II(S)$ is a subset of O and that $IE(S)$ is a subset of $\overline{O}$ and also that any π-path from an element of O to an element (necessarily not the same) of $\overline{O}$ must cross S. Now suppose that c is in O. Then any π-path from c to an element of $IE(S)$ must cross S, showing that c is not in $E(S)$, which means that (3.2.7) is true. Now suppose that d is in $\overline{O}$. Then any π-path from d to an element of $II(S)$ must cross S, showing that d is not in $I(S)$, which means that (3.2.8) is true. Thus, (ii) implies (iii) and the O of (ii) has to be $I(S)$. Finally, (iii) of this theorem implies (iii) of Lemma 3.2.2, which is shown there to be equivalent to (i). $\square$

3.3. Connectedness in Digital Spaces

One of the general aims of this book is to find an acceptable definition of the notion of a "Jordan surface" in a digital space. In view of the results of the previous section, it is reasonable to look for a definition which says something like "a surface is Jordan if it is near-Jordan and its interior and its exterior are both connected in some sense." The purpose of the current section is to make precise the sense of this connectedness.

First we give a simple example (see Figure 3.3.1) to show that there is a near-Jordan surface in $\left(Z^2, w_2\right)$ for which neither the interior nor the exterior is w_2-connected. We chose the commonly used digital space $\left(Z^2, w_2\right)$ (i.e., a square grid with edge-adjacency) for this example to make it clear that the phenomenon illustrated by it is not due to the existence of some peculiar digital space allowed by our definitions.

One way to ensure that Jordan surfaces have the sort of property deemed desirable in the first paragraph of this section is to define a surface in (V, π) as Jordan if it is near-Jordan and both its interior and its exterior are π-connected. However, this would be too restrictive, since it would rule out many cases already found useful in the literature. An example is provided by Figure 3.3.2. For reasons explained in Chapter 1, it is standard practice to allow different adjacencies to determine connectedness for the interior and for the exterior of surfaces. Using 8-adjacency (α_2) for the interior and 4-adjacency (w_2) for the exterior results in the surface of Figure 3.3.2 having both a connected interior and a connected exterior, and thus it would be considered "Jordan." This is how it would be viewed by the standard approach in the literature. If we insisted on using 4-adjacency for both the interior and the exterior, then the surface would not be considered "Jordan." In what follows we adopt the approach of using different adjacencies for the interior and the exterior in our more general environment. (Another relevant reason for using different adjacencies for the interior and the exterior is that the resulting boundaries may be easier to track than if we used a pair of identical adjacencies, even though this pair could not be rejected as unsuitable by arguments based purely on Jordan surface properties; see Chapter 8.)

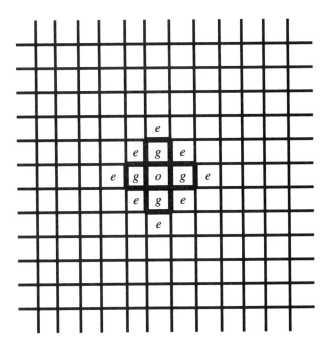

Figure 3.3.1. Example of a near-Jordan surface in (Z^2, ω_2) for which neither the interior nor the exterior is ω_2-connected. The bold edges correspond to surfels in S. For each such surfel, the first element is the spel labeled g, and the second element is the spel on the other side of the corresponding edge. In this case, $I(S) = II(S)$ consists of the four spels labeled g. Elements of $IE(S)$ are o and the eight spels labeled e. All unlabeled spels are in the exterior of S. Clearly, neither $I(S)$ nor $E(S)$ is ω_2-connected. (Reproduced from [14] with the publisher's permission.)

We call a symmetric binary relation ρ on V a *spel-adjacency* in the digital space (V, π) if $\pi \subset \rho$. Trivially, π and V^2 are spel-adjacencies in (V, π). The binary relations ω_2 and α_2 are spel-adjacencies in both (Z^2, χ) and (Z^2, ω_2), as are the binary relations β, γ, and δ on Z^2 which are illustrated in Figure 2.2.1. A further spel-adjacency in both (Z^2, χ) and (Z^2, ω_2) is given by

$$\nu = \left\{ \, ((c_1, c_2), (d_1, d_2)) \mid [\, c_2 = d_2 \,\, \& \,\, c_1 \times d_1 \geq 0 \,] \text{ or} \atop [\, |c_2 - d_2| = 1 \,\, \& \,\, c_1 = d_1 \,] \, \right\}, \tag{3.3.1}$$

which is illustrated in Figure 3.3.3.

There is an interesting sense in which the spel-adjacency ν is different from the others, and now we take a little diversion to discuss this. As in Section 1.4, let M be any set and ρ be a binary relation on M. We say that ρ is a *finitary binary relation on M* if, for every $c \in M$, the set $\{\, d \in M \mid (c, d) \in \rho \,\}$ is finite. It is clear from the definitions that ω_2, α_2, β, γ, and δ are finitary binary relations on Z^2, but ν is not.

From the definitions of a digital space and of a spel-adjacency, it immediately follows that if (V, π) is a digital space and ρ is a spel-adjacency in (V, π), then (V, ρ) is a digital space. We say that a digital space (V, π) is a *finitary digital space* if π is a finitary binary

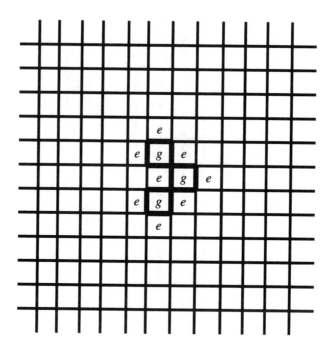

Figure 3.3.2. Example of a near-Jordan surface in $\left(Z^2, \omega_2\right)$ for which the interior is not ω_2-connected but is α_2-connected, and the exterior is ω_2-connected. The bold edges correspond to surfels in S. For each such surfel, the first element is the spel labeled g and the second element is the spel labeled e. In this case, $I(S) = II(S)$ consists of the three spels labeled g. Elements of $IE(S)$ are the eight spels labeled e. All unlabeled spels are in the exterior of S. (Reproduced from [16] with the publisher's permission.)

relation on V. In view of this definition, $\left(Z^2, \omega_2\right)$, $\left(Z^2, \alpha_2\right)$, $\left(Z^2, \beta\right)$, $\left(Z^2, \gamma\right)$, and $\left(Z^2, \delta\right)$ are all finitary digital spaces, but $\left(Z^2, \nu\right)$ is not.

An interesting aspect of the theory that we are developing is that, generally speaking, basic results about digital spaces do **not** require the assumption that they be finitary. We have already seen this in Lemmas 3.2.1 and 3.2.2 and in Theorem 3.2.5. As we will see, it is essentially the case that we will need finitariness only when we wish to prove either that something or other is finite or that an algorithm terminates in a finite number of steps.

Now we give the definition which formalizes our intuitive idea of a "Jordan surface." Let κ and λ be arbitrary spel-adjacencies in a digital space (V, π). A surface in (V, π) is said to be a $\kappa\lambda$-*Jordan surface* if it is near-Jordan, its interior is κ-connected, and its exterior is λ-connected.

The surface in (Z^2, χ) illustrated in Figure 3.2.1 is near-Jordan, but it is not $\chi\chi$-Jordan, since its exterior is not χ-connected. On the other hand, it is $\chi\nu$-Jordan, since ν connects pixels which are one above another; see Figure 3.3.3. The surface in Figure 3.2.2 is not near-Jordan, and hence it is not $\kappa\lambda$-Jordan for any spel-adjacencies κ and λ in (Z^2, χ). The near-Jordan surface in $\left(Z^2, \omega_2\right)$ illustrated in Figure 3.3.1 is not $\omega_2\omega_2$-Jordan, $\omega_2\alpha_2$-Jordan, or $\alpha_2\omega_2$-Jordan, but it is $\alpha_2\alpha_2$-Jordan. The near-Jordan surface in $\left(Z^2, \omega_2\right)$ illustrated in Figure 3.3.2 is not $\omega_2\omega_2$-Jordan or $\omega_2\alpha_2$-Jordan, but it is $\alpha_2\omega_2$-Jordan and $\alpha_2\alpha_2$-Jordan.

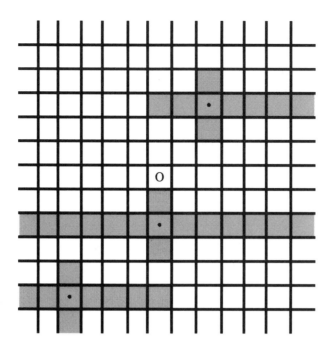

Figure 3.3.3. Illustration of the spel-adjacency ν. The origin is labeled O. The spels adjacent to the spels labeled • are shaded.

Theorem 3.3.1. *Let κ and λ be spel-adjacencies in a digital space (V, π). A surface S in (V, π) is $\kappa\lambda$-Jordan if, and only if, there exists a nonempty κ-connected subset O of V such that $\overline{O}$ is nonempty and λ-connected and $S = \partial(O, \overline{O})$. Furthermore, for such an O, $I(S) = O$ and $E(S) = \overline{O}$.*

Proof. If S is a $\kappa\lambda$-Jordan surface, then $I(S)$ is κ-connected and $E(S)$ is λ-connected. Letting $O = I(S)$, we see from Lemmas 3.2.1 and 3.2.2 that $\overline{O} = E(S)$ and $S = \partial(O, \overline{O})$.

Conversely, suppose that O is a nonempty κ-connected subset such that $\overline{O}$ is nonempty and λ-connected and $S = \partial(O, \overline{O})$. It follows from Theorem 3.2.5 that S is a $\kappa\lambda$-Jordan surface with $I(S) = O$ and $E(S) = \overline{O}$. $\square$

This theorem gives us a characterization of $\kappa\lambda$-Jordan surfaces. However, it is not a very useful characterization, since even in the special case of (Z^3, ω_3) it is not particularly easy to check on its validity as we attempt to form a surface based on information assigned to individual spels. On the other hand, the following is a useful tool for proving connectedness of the interiors and/or the exteriors of surfaces.

Lemma 3.3.2. *Let S be a surface, and let κ and λ be spel-adjacencies in a digital space (V, π).*

(i) *If for some κ-connected subset O of V*

$$II(S) \subset O \subset I(S),$$ (3.3.2)

then $I(S)$ is κ-connected.

(ii) *If for some λ-connected subset Q of V*

$$IE(S) \subset Q \subset E(S),\tag{3.3.3}$$

then $E(S)$ is λ-connected.

Proof. We prove only (i) since the proof of (ii) is entirely analogous. We assume the truth of the premise of (i). Let c_1 and c_2 be two spels in $I(S)$. We need to prove that they are κ-connected in $I(S)$. By (3.2.2), for $1 \leq i \leq 2$, there exists a d_i in $II(S)$ such that there is a π-path (hence, by the definition of a spel-adjacency, a κ-path) not crossing S (and hence entirely in $I(S)$) connecting c_i to d_i. By (3.3.2), d_1 and d_2 are in O and are therefore κ-connected in O and hence (again by (3.3.2)) in $I(S)$. By combining these three κ-paths in $I(S)$ into a single κ-path (recall that κ is symmetric by the definition of a spel-adjacency), we get the desired result. $\square$

Note that the proof of this lemma made essential use of both conditions for a binary relation to be a spel-adjacency, namely, that it must contain π and that it must be symmetric.

Since the following theorem is one of our main results, before stating it in a mathematically concise way, we give a detailed description of an interesting special case of it. For an arbitrary spel-adjacency κ, we can start with any nonempty κ-connected proper subset O of V. We consider that this is an "object" in the digital space. Any spel-adjacency λ partitions the set of spels not in the object into components. Let Q be one of these components, and let S be the surface between O and Q. Then the exterior of S is exactly Q. Its interior consists of all the spels not in Q (and hence, in particular, contains all the spels in O) and is κ-connected. Furthermore, one cannot get from any spel not in Q (and hence from any spel in O) to a spel in Q without crossing S.

Theorem 3.3.3. *Let κ and λ be spel-adjacencies in a digital space (V, π). Let O be a nonempty κ-connected proper subset of V, and let Q be a λ-connected union of π-components of $\overline{O}$. (In particular, any one of the λ-components of $\overline{O}$ is a valid choice for Q.) Then $S = \partial(O, Q)$ is a $\kappa\lambda$-Jordan surface. Furthermore, $E(S) = Q$ and $I(S) = \overline{Q}$.*

Proof. First we observe that the hypotheses imply that

$$S = \partial(O, Q) = \partial(\overline{Q}, Q).\tag{3.3.4}$$

(The first boundary is a subset of the second one, since $O \subset \overline{Q}$. On the other hand, if (c, d) is in $\partial(\overline{Q}, Q)$, then c must be in O; and hence (c, d) must be in $\partial(O, Q)$. Otherwise, the π-adjacency of c and d would imply that they are in the same π-component of $\overline{O}$, contradicting the facts that d is in Q but c is not.) Therefore it follows from Theorem 3.2.5 that S is near-Jordan and that $I(S) = \overline{Q}$ and $E(S) = Q$. To complete the proof, all we need to show is that $I(S)$ is κ-connected. This follows from Lemma 3.3.2(i), since (recalling again that $O \subset \overline{Q}$) we see that (3.3.2) holds. $\square$

As an illustration of this theorem, consider Figure 3.3.1. Let O be the α_2-connected set consisting of the four spels labeled g. If we choose Q as the ω_2-component of $\overline{O}$ which consists of all the spels labeled e and all the unlabeled spels, then the resulting surface consists of the twelve surfels which are pairs consisting of a spel labeled g and a spel labeled e. The interior of this surface consists of the spels labeled g (that is, spels in O) and the spel labeled o. The exterior of the surface is Q, as required by the theorem. If instead we choose Q as the ω_2-component of $\overline{O}$ which consists of the spel labeled o, then the resulting surface consists of the surfels which are pairs of a spel labeled g and the spel labeled o. The interior of this surface consists of the spels labeled g (that is, spels in O), the spels labeled e, and all the unlabeled spels. The exterior of the surface consists of the spel labeled o (that is, the only spel in Q). Finally, if we choose Q as the α_2-component consisting of all the spels not in O, then the resulting surface consists of all surfels associated with bold edges in Figure 3.3.1, and its interior and exterior are just O and Q ($= \overline{O}$), respectively.

We have started this section with a discussion of the fact that near-Jordanness of a surface does not guarantee the connectedness either of its interior or of its exterior; see Figure 3.3.1. Now we are going to introduce one of the most interesting and important notions of this book: a special type of near-Jordan surface for which the connectedness of the interior and the connectedness of the exterior are automatically satisfied.

First, let us motivate our definition by analogy to the continuous case. Look at Figure 1.5.1. We see that the simple closed curves illustrated in that figure have a property not shared by near-Jordan surfaces: if we remove any nonempty proper subset of such a curve, the set of points left behind no longer separates an interior set of points from an exterior set of points. (The original topologically connected interior can be connected to the original topologically connected exterior across the removed part of the original curve.) This minimality property of simple closed curves is not shared by near-Jordan surfaces. For example, the near-Jordan surface S defined in the caption of Figure 3.3.1 consists of sixteen surfels. If we remove from S the four surfels surrounding the left-most g in Figure 3.3.1, then the twelve surfels which remain form the near-Jordan surface of Figure 3.3.2.

As a way of capturing the minimality property of simple closed curves, we say that S is a *minimally near-Jordan surface* in a digital space (V, π) if S is a near-Jordan surface in (V, π) such that any surface in (V, π) which is a proper subset of S is not near-Jordan. The truly fascinating thing is that minimality guarantees the connectedness of the interior and of the exterior, as we now proceed to prove.

Theorem 3.3.4. *If S is a minimally near-Jordan surface in a digital space (V, π), then both $I(S)$ and $E(S)$ are π-connected.*

Proof. We prove the result only for $E(S)$. The proof for $I(S)$ is strictly analogous. Since S is near-Jordan, it follows from Lemmas 3.2.1 and 3.2.2 that $E(S)$ is the complement of $I(S)$ and $S = \partial(I(S), E(S))$. We can always find a spel-adjacency κ in (V, π) such that $I(S)$ is κ-connected. (Recall that even V^2 is a spel-adjacency.) Also, since S is a surface, neither $I(S)$ nor $E(S)$ is empty, and so $I(S)$ is a nonempty proper subset of V. To complete the proof, now we assume that $E(S)$ is not π-connected and show that this leads to a contradiction.

Let Q' and Q'' be two distinct (and hence disjoint) π-components of $E(S)$. Then, by Theorem 3.3.3, both $S = \partial(I(S), Q')$ and $S = \partial(I(S), Q'')$ are $\kappa\pi$-Jordan surfaces, which in particular means that they are nonempty and near-Jordan. They are also disjoint, implying that they are both near-Jordan surfaces which are proper subsets of S, contradicting the minimal near-Jordanness of S. □

Theorem 3.3.5.   *If S is a surface in a digital space (V, π), then S is minimally near-Jordan if, and only if, it is $\pi\pi$-Jordan.*

Proof. It is an immediate consequence of Theorem 3.3.4 and the definitions that if S is minimally near-Jordan, then it is $\pi\pi$-Jordan.

Conversely, assume that S is $\pi\pi$-Jordan. By definition, this implies that S is near-Jordan. Let S' be a surface in (V, π) which is a proper subset of S. Our proof is complete if we can show that S' is not near-Jordan.

Since S' is a surface, it is nonempty and hence it contains a surfel (c, d). Since S' is a subset of S, also $(c, d) \in S$. Since S' is a proper subset of S, there must also be a surfel (e, f) which is in S but not in S'. Since S is $\pi\pi$-Jordan, there is a π-path $\langle c^{(0)}, \cdots, c^{(K)} \rangle$ in $I(S)$ from c to e. This path does not cross S (otherwise it would also contain an element of $E(S)$, which is impossible for a near-Jordan S, see Lemma 3.2.2), and hence it does not cross its subset S'. Similarly, there is a π-path $\langle d^{(0)}, \cdots, d^{(L)} \rangle$ in $E(S)$ from f to d which does not cross S'. Then the π-path $\langle c^{(0)}, \cdots, c^{(K)}, d^{(0)}, \cdots, d^{(L)} \rangle$ is from the spel c in $II(S')$ to the spel d in $IE(S')$ and does not cross S'. (Recall that $(c^{(K)}, d^{(0)}) = (e, f) \in S - S'$.) This proves that S' is not near-Jordan. $\square$

Rather than using two alternative names for the concepts which have been proven to be the same in this theorem, we will call S a *Jordan surface* in a digital space (V, π) if it is a minimally near-Jordan surface or, equivalently, a $\pi\pi$-Jordan surface in (V, π). This terminology is more than justified by the following consequence of Lemmas 3.2.1 and 3.2.2 combined with Theorem 3.3.4.

Corollary 3.3.6.   *A Jordan surface S in a digital space (V, π) has the following properties.*

(i) $S = \partial(I(S), E(S))$.

(ii) $I(S) \cup E(S) = V$ and $I(S) \cap E(S) = \emptyset$.

(iii) *Both $I(S)$ and $E(S)$ are π-connected.*

(iv) *Every π-path from any element of $I(S)$ to any element of $E(S)$ crosses S.*

We complete this section by introducing the notion of the "reverse-oriented version" of a surface. For any surface S in a digital space, we define the surface $\widetilde{S}$ in the same space by

$$\widetilde{S} = \{ (d, c) \mid (c, d) \in S \} . \tag{3.3.5}$$

Theorem 3.3.7.   *Let S be a surface in a digital space (V, π).*

(i) $\widetilde{\widetilde{S}} = S$.

(ii) $S = \partial(O, Q)$ *if, and only if, $\widetilde{S} = \partial(Q, O)$.*

(iii) $II(\widetilde{S}) = IE(S)$ *and* $IE(\widetilde{S}) = II(S)$.

(iv) $I(\widetilde{S}) = E(S)$ *and* $E(\widetilde{S}) = I(S)$.

(v) S *is near-Jordan if, and only if, $\widetilde{S}$ is near-Jordan.*

(vi) *If κ and λ are spel-adjacencies in (V, π), then S is a $\kappa\lambda$-Jordan surface in (V, π) if, and only if, $\widetilde{S}$ is a $\lambda\kappa$-Jordan surface in (V, π).*

(vii) S *is a Jordan surface in (V, π) if, and only if, $\widetilde{S}$ is a Jordan surface in (V, π).*

Proof. These results are straightforward consequences of the definitions. For example, (iv) follows from (iii) in view of (3.2.2) and the fact that a π-path crosses S if, and only if, it crosses $\widetilde{S}$. Then (v) follows from (iv), due to the necessary and sufficient condition for a surface to be near-Jordan that is expressed in Lemma 3.2.2(iii). Finally, (iv) and (v) immediately imply (vi), which in turn immediately implies (vii). $\square$

3.4. Isomorphisms between Digital Spaces

Until now we have been using the notion of isomorphism between digital spaces on the sly. For example, we have interchangeably used the digital space whose set of spels is Z^2 (i.e., ordered pairs of integers) and whose set of surfels is ω_2 and the digital space whose set of spels is a set of points in the plane arranged in a square grid and whose set of surfels are those pairs of grid points whose Voronoi neighborhoods share exactly an edge. If we give a strictly algebraic interpretation to the first space and a strictly geometrical interpretation to the second space, then these spaces are **not** the same. The way we handled them is justified by the fact that they are isomorphic (in the sense just to be defined) and, as a consequence, whatever can be said within the theory of digital spaces about the one can also be said about the other after a suitable translation of the terminology.

An *isomorphism* i from a digital space (V, π) to a digital space (V', π') is a one-to-one function which maps V onto V' so that

$$(c, d) \in \pi \iff (i(c), i(d)) \in \pi'. \tag{3.4.1}$$

(Recall that *one-to-one* means that $i(c) = i(d)$ implies that $c = d$ and *onto* means that for every $c' \in V'$ there is a $c \in V$ such that $i(c) = c'$. Such a one-to-one onto function i always has an *inverse* i^{-1} such that $i^{-1}(i(c)) = c$ for all $c \in V$ and $i(i^{-1}(c')) = c'$ for all $c' \in V'$.) Clearly, the inverse of i in such a case is an isomorphism from (V', π') to (V, π). This justifies using the terminology that there is an *isomorphism between* (V, π) and (V', π') or that they are *isomorphic digital spaces*.

Now we illustrate the claim made above that, whenever two digital spaces are isomorphic, "whatever can be said within the theory of digital spaces about the one can also be said about the other after a suitable translation of the terminology" on a simple example. The validity of this general claim justifies not distinguishing between isomorphic digital spaces. Within the confines of our theory they can indeed be considered the "same."

Theorem 3.4.1. *Let i be an isomorphism from a digital space (V, π) to a digital space (V', π'), and let S be a surface in (V, π). If we define*

$$S' = \{ (i(c), i(d)) \mid (c, d) \in S \}, \tag{3.4.2}$$

then S is a near-Jordan surface in (V, π) if, and only if, S' is a near-Jordan surface in (V', π').

Proof. First we point out that S' is indeed a surface in (V', π'), since the fact that S is a nonempty subset of π implies, in view of (3.4.1), that S' is a nonempty subset of π'.

The rest of the theorem is proved by the following sequence of equivalences, each one of which is an immediate consequence of the definitions. S is not a near-Jordan surface in (V, π) if, and only if, there exists a π-path $\langle c^{(0)}, \cdots, c^{(K)} \rangle$ from an element of $II(S)$ to an element of $IE(S)$ which does not cross S. This happens if, and only if, there exists a π'-path $\langle i(c^{(0)}), \cdots, i(c^{(K)}) \rangle$ from an element of $II(S')$ to an element of $IE(S')$ that does not cross S', which means that S' is not a near-Jordan surface in (V', π'). $\square$

A trivial, but important, class of isomorphisms contains those due to "scaling" in euclidean spaces. For any digital space (V, π) for which V is a grid in a euclidean space R^N and for any positive real number σ, let $V' = \{ \sigma c \mid c \in V \}$ and let $\pi' = \{ (\sigma c, \sigma d) \mid (c, d) \in \pi \}$. Clearly, the function i defined by $i(c) = \sigma c$ is an isomorphism from (V, π) to (V', π'). In particular, this means that theorems proven about the grid Z^N can be immediately translated into theorems about the grid δZ^N, for any positive δ. We will not bother to do this but will deal with the grid Z^N only. For this reason, from now on we refer to Z^2 as *the* square grid and to Z^3 as *the* cubic grid. Similarly, we will refer to $F = F_1$ as *the* fcc grid and to $B = B_1$ as *the* bcc grid.

For our next example we work out the proof of the isomorphism in detail, even though geometric intuition should make us quite certain that the specified function is indeed an isomorphism. We do this since we consider it instructive to present a simple formal proof based on the algebraic definitions.

Theorem 3.4.2. *The digital spaces $\left(Z^2, \beta \right)$ and $\left(Z^2, \gamma \right)$ are isomorphic.*

Proof. Define, for $c \in Z^2$, $i(c) = (c_1, -c_2)$. Now we are going to show that i is an isomorphism from $\left(Z^2, \beta \right)$ to $\left(Z^2, \gamma \right)$. Clearly, i is a one-to-one function which maps Z^2 onto itself. By looking at (3.4.1) for this special case, we see that it is satisfied since the conditions

βi. $|c_1 - d_1| \leq 1$
βii. $|c_2 - d_2| \leq 1$
βiii. $(c_1 - d_1) \neq (d_2 - c_2)$

are satisfied if, and only if, the conditions

γi. $|c_1 - d_1| \leq 1$
γii. $|-c_2 + d_2| \leq 1$
γiii. $(c_1 - d_1) \neq (-c_2 + d_2)$

are satisfied. $\square$

For a more interesting example, consider the digital space $\left(Z^2, \beta \right)$ and the digital space (H, ε), where H is the hexagonal grid and ε is the edge-adjacency between the hexagons which are the Voronoi neighborhoods of the grid points (see Figure 2.2.3). In this case, the isomorphism is defined by

$$i(c) = c_1 \times (1, 0) + c_2 \times (-0.5, \sqrt{0.75}), \qquad (3.4.3)$$

and it is easy to see, on the basis of the figures representing the respective proto-adjacencies, that this is indeed an isomorphism from $\left(Z^2, \beta \right)$ to (H, ε).

Quite obviously (at least we hope that it is obvious), if (V, π) is isomorphic to (V', π') and (V', π') is isomorphic to (V'', π''), then (V, π) is isomorphic to (V'', π''). Hence $\left(Z^2, \beta \right)$ and $\left(Z^2, \gamma \right)$ can be thought of as two alternative (both somewhat asymmetrical) formal ways of representing the (very symmetrical) hexagonal grid with edge-adjacency. Now, why would we want to do such a thing? Well, it turns out that this is useful route to proving important theorems about grids such as the hexagonal grid. Just hang in there, and you will see.

In fact, this seems like a good place to take a big leap and generalize what we have been discussing from dimension 2 to an arbitrary dimension $N \geq 2$. An *N-dimensional sign function s* assigns to every unordered pair of integers $\{i, j\}$ $(1 \leq i \neq j \leq N)$ a value $s_{\{i,j\}}$ such that $|s_{\{i,j\}}| = 1$. For each such sign function s, we consider the spel-adjacency β_s in (Z^N, ω_N) which is defined as follows. In addition to all elements of ω_N, β_s contains a pair of spels (c, d) if there exist integers i and j $(1 \leq i \neq j \leq N)$ such that

$$(c_i - d_i) \times (c_j - d_j) = s_{\{i,j\}}, \tag{3.4.4}$$

and $c_n = d_n$ if n is neither i nor j.

When $N = 2$, there is only one possible way to choose the $\{i, j\}$ of the previous paragraph: it has to be that $\{i, j\} = \{1, 2\}$. There are two possible values of $s_{\{1,2\}}$, namely, 1 and −1. It is easy to check that in the first case $\beta_s = \beta$ and in the second case $\beta_s = \gamma$. Thus, in either case, (Z^2, β_s) is isomorphic to the hexagonal grid with edge-adjacency (equivalently, with edge-or-vertex-adjacency).

When $N = 3$, there are eight possible choices for β_s, and each of the eight digital spaces (Z^3, β_s) is isomorphic to (F, β_1), i.e., to the fcc grid with face-adjacency (equivalently, face-or-edge-adjacency). Here we prove this for one of the choices for s, the one illustrated in Figure 3.4.1.

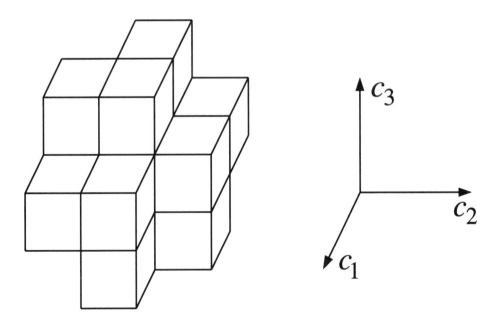

Figure 3.4.1. Illustration of the adjacency $\beta_{\bar{s}}$ discussed in Theorem 3.4.3. The (not visible) central voxel has twelve voxels $\beta_{\bar{s}}$-adjacent to it. Nine of these are visible in the figure, and the other three are to the left, below, and behind, respectively, the central voxel. The relationship between this adjacency and the face-adjacency β_1 for the fcc grid (illustrated in Figure 2.1.2) is a three-dimensional version of the relationship between the adjacency γ illustrated in Figure 2.2.1 and the edge-adjacency for the hexagonal grid (illustrated in Figure 2.2.3).

Theorem 3.4.3. *Let $\bar{s}$ be the three-dimensional sign function whose value is always* -1. *The digital spaces* $(Z^3, \beta_{\bar{s}})$ *and* (F, β_1) *are isomorphic.*

Proof. Define, for $c \in Z^3$, $i(c) = (c_2 + c_3, c_3 + c_1, c_1 + c_2)$. Since the sum of its components is even, $i(c) \in F$. To show that i is one-to-one, assume that $i(c') = i(c'')$, and let $c = c' - c''$. Then $i(c) = i(c') - i(c'') = 0$, which implies that $c_2 = -c_3$, $c_1 = -c_3$, and $-2c_3 = c_1 + c_2 = 0$. From this it follows that all three components of c are 0, and hence $c' = c''$. Finally, to show that i maps Z^3 onto F, let $d \in F$, and define $c_1 = (-d_1 + d_2 + d_3)/2$, $c_2 = (d_1 - d_2 + d_3)/2$, and $c_3 = (d_1 + d_2 - d_3)/2$. Since the sum of the components of d is even, c_1, c_2, and c_3 are all integers, and, clearly, $i(c) = d$. This shows that i is a one-to-one function mapping Z^3 onto F.

To show that i is an isomorphism from $(Z^3, \beta_{\bar{s}})$ to (F, β_1), first we observe that $(i(c), i(d)) \in \beta_1$ if, and only if, $i(c)$ and $i(d)$ differ by 1 in two coordinates and are the same in the third. A consequence of this observation is that if we let $a = i(c) - i(d)$, then $(i(c), i(d)) \in \beta_1$ if, and only if, two of $|a_1|$, $|a_2|$, and $|a_3|$ have the value 1 and the third has the value 0. By the definition of i, for $a = i(c) - i(d)$,

$$a_1 = (c_2 - d_2) + (c_3 - d_3),$$
$$a_2 = (c_3 - d_3) + (c_1 - d_1), \qquad (3.4.5)$$
$$a_3 = (c_1 - d_1) + (c_2 - d_2).$$

Now suppose that $(c, d) \in \beta_{\bar{s}}$. There are two different reasons why this might be the case. The first is that $(c, d) \in \omega_3$. In this case, exactly one of $|c_1 - d_1|$, $|c_2 - d_2|$, and $|c_3 - d_3|$ has the value 1 and the other two have the value 0. The other possibility is that there exist two integers k and j ($1 \le k \ne j \le 3$) such that $(c_k - d_k) \times (c_j - d_j) = -1$ and $c_n = d_n$ if n is neither k nor j. Looking at the components of a, as expressed in the equation above, we see that either case implies that $(i(c), i(d)) \in \beta_1$.

Conversely, suppose that $(i(c), i(d)) \in \beta_1$ and so two of $|a_1|$, $|a_2|$, and $|a_3|$ have the value 1 and the third has the value 0. One way that this may come about is by both the terms in the zero-valued sum being themselves zero-valued. In that case, the other (common) term in the other two sums must have absolute value 1. This implies that $(c, d) \in \omega_3$. The alternative possibility is that the terms in the zero-valued sum are not zero-valued but are the negatives of each other. Then we have the situation that there exist two integers k and j ($1 \le k \ne j \le 3$) such that $(c_k - d_k) = -(c_j - d_j) \ne 0$ and, if n is neither k nor j, $|(c_n - d_n) + (c_k - d_k)| = 1$ and $|(c_n - d_n) - (c_k - d_k)| = 1$. It is easy to see that this can happen only if $(c_k - d_k) \times (c_j - d_j) = -1$ and $c_n = d_n$. In either case, $(c, d) \in \beta_{\bar{s}}$. $\square$

Since, as we have just discussed, for any two-dimensional sign function s, (Z^2, β_s) is isomorphic to the hexagonal grid with edge-adjacency and, for any three-dimensional sign function s, (Z^3, β_s) is isomorphic to the fcc grid with face-adjacency, now we see a justification of the claim made in Chapter 2 that the appropriate two-dimensional analog of the fcc grid is the hexagonal grid. Now we turn to the validity of the same claim but with the bcc grid in place of the fcc grid.

For any $N \ge 1$, let e be an element of Z^N such that $e_1 = 1$ and, for $2 \le n \le N$, $|e_n| = 1$. We call such an e an *N-dimensional direction vector*, and we associate with it the spel-adjacency ε_e in (Z^N, ω_N) which is defined by: $(c, d) \in \varepsilon_e$ if, and only if, $c \ne d$ and at least one of the following two conditions is satisfied:

    (i) for $1 \le n \le N$, $c_n - d_n$ is either e_n or 0;

    (ii) for $1 \le n \le N$, $c_n - d_n$ is either $-e_n$ or 0.

There is only one one-dimensional direction vector, namely, $e = (1)$. Obviously, in this one-dimensional case, $\varepsilon_e = \omega_1$.

There are two two-dimensional direction vectors $(1, 1)$ and $(1, -1)$. We leave it to the reader to justify that, if e is the first of these, then $\varepsilon_e = \beta$ and if e is the second of these, then $\varepsilon_e = \gamma$. It follows that, for any two-dimensional direction vector e, (Z^2, ε_e) is isomorphic to the digital space defined by the hexagonal grid with edge-adjacency.

There are four three-dimensional direction vectors e and, for each one of these, the resulting digital space (Z^3, ε_e) is isomorphic to the digital space defined by the bcc grid with face-adjacency (equivalently, face-or-edge-adjacency or face-or-edge-or-vertex adjacency; see Exercise 2.3). Here we indicate only the proof of the isomorphism between $(Z^3, \varepsilon_{(1,1,1)})$ and the bcc grid in which adjacency is defined as follows: two spels of the bcc grid are adjacent if either they differ only in one of their components and the absolute value of the difference in that component is 2 or they differ in all three components and the absolute values of the differences in each of the components is 1 (see Exercise 2.4). The function defined by $i(c) = (-c_1 + c_2 + c_3, \; c_1 - c_2 + c_3, \; c_1 + c_2 - c_3)$ is a one-to-one mapping of Z^3 onto B. We leave it to the reader to check the details of the claim that, for any c and d in Z^3, $(c, d) \in \varepsilon_{(1,1,1)}$ if, and only if, $i(c)$ and $i(d)$ are adjacent in the bcc grid.

We complete this section with the discussion of a notion prevalent in the literature. A digital space (V, π) is said to be *translation-invariant* if V is a grid in R^N (for some positive integer N) and for any c and d in V, the function t_{c-d} defined by

$$t_{c-d}(e) = e + c - d \tag{3.4.6}$$

(for all $e \in V$) is an isomorphism from (V, π) to (V, π). Most, but not all, digital spaces that have been studied in the literature are translation-invariant. A notable exception is Z^2 with the Khalimsky adjacency, as depicted in Figure 2.2.5. A more artificial example is (Z^2, ν) illustrated in Figure 3.3.3.

It turns out that translation invariance is not an important property from the point of view of our theory. None of our important theorems need to assume that the underlying digital space is translation-invariant. Where translation invariance is useful is in the design of algorithms. If a digital space is translation-invariant, what needs to be done at any place in it does not depend on where that place is. Rather, it depends only on what is at that place and in its neighborhood. Such localized behavior of algorithms was already demonstrated in our first chapter. The various algorithms for flat, fat, and cloning flies used only the local configuration of the sugar cubes in controlling the behavior of the flies. This was possible, since (Z^N, ω_N) is translation-invariant. To a fly a particular local arrangement of sugar cubes looks the same, wherever it is.

It is worth reemphasizing the nature of isomorphic digital spaces: they are the "same" both from the mathematical and from the computational points of view. Mathematically, theorems about one of them translate into theorems about the other (and vice versa). Computationally, an algorithm which achieves its aim in one space will achieve the corresponding aim in the space isomorphic to it. We are thus justified in treating isomorphic spaces as if they were the "same." Thus, the exact shape of the voxels makes no difference whatsoever to Artzy's algorithm; as long as the digital space is isomorphic to a cubic grid, Artzy's algorithm will have all the desirable properties that we have claimed for it in Z^3.

3.5. Exercises

3.1. A π-path $\langle c^{(0)}, \cdots, c^{(K)} \rangle$ in a digital space (V, π) is said to be *exiting through* a surface S if there is a k, $1 \leq k \leq K$, such that $(c^{(k-1)}, c^{(k)}) \in S$. Let

$$I'(S) = \{ c \in V \mid \text{there exists a } \pi\text{-path, connecting an element}$$
$$\text{of } II(S) \text{ to } c, \text{ which is not exiting through } S \}, \text{ and}$$
$$E'(S) = \{ c \in V \mid \text{there exists a } \pi\text{-path, connecting } c \text{ to an}$$
$$\text{element of } IE(S), \text{ which is not exiting through } S \}. \tag{3.5.1}$$

(Such definitions were used, for example in [16].) Prove that, for any surface S in a digital space (V, π),

$$I(S) = I'(S) \text{ and } E(S) = E'(S). \tag{3.5.2}$$

3.2. Using the definition of "exiting through" given in the previous exercise, prove that, for any surface S in any digital space (V, π), S is near-Jordan if, and only if, every π-path from any element of $II(S)$ to any element of $IE(S)$ is exiting through S.

3.3. Show that the ϕ defined on page 42 is a spel-adjacency in (Z^4, ω_4).

3.4. Is it true for the set $O = \{ (z, z, 0, z) \mid z \in Z \}$ of spels in (Z^4, ω_4) that $II(\partial(O, \overline{O}))$ is ϕ-connected but not ω_4-connected?

3.5. Is there an ω_4-path $\langle c^{(0)}, \cdots, c^{(K)} \rangle$ from $(0, 0, 0, 0)$ to $(1, 1, 1, 0)$ such that, for $1 \leq k \leq K$, $(c^{(0)}, c^{(k)}) \in \phi$ and $c_4^{(k)} = 0$?

3.6. Prove that either of the following two conditions is sufficient to make S an antisymmetric surface in (V, π).
(i) S is a nonempty boundary between disjoint subsets of V.
(ii) S is a near-Jordan surface in (V, π).

3.7. Define a *semidigital space* as a pair (V, π), where V is an arbitrary nonempty set and π is a binary relation (not necessarily symmetric) on V such that V is π-connected, and define a *surface* in such a space as any nonempty subset of π. We can use (3.2.1), (3.2.2), and (3.5.1) to define $I(S)$, $E(S)$, $I'(S)$, and $E'(S)$. Show that Lemma 3.2.1 holds even for semidigital spaces, but that it is **not** necessarily the case that, for any surface S in a semidigital space (V, π),

$$I'(S) \cup E'(S) = V. \tag{3.5.3}$$

(Hint: a possible counterexample is in (Z^2, ψ), with

$$\psi = \{ (c, d) \mid [\ |c_1 - d_1| = 1 \ \& \ c_2 = d_2 \neq 0 \ ]$$
$$\text{or } [\ c_1 - d_1 = 1 \ \& \ c_2 = d_2 = 0 \ ] \tag{3.5.4}$$
$$\text{or } [\ |c_2 - d_2| = 1 \ \& \ (c_1 = d_1 = 10 \times k, \text{ for some integer } k) \ ] \}.)$$

3.8. Prove that (Z^1, ω_1) is not a boundable digital space.

3.9. Let G be a Delone grid in R^2, and let π be defined by the edge-adjacency of the polygons which are the Voronoi neighborhoods of the grid points (see Exercise 1.5). Prove that (G, π) is a boundable digital space. (It follows from Theorem 3.2.4 that, for every finite near-Jordan surface S in (G, π), exactly one of $I(S)$ or $E(S)$ is finite.)

3.10. Prove that the following are finitary digital spaces (all the notation has been defined in Chapters 1 and 2):
(i) for any positive integer N, (Z^N, α_N);
(ii) for any positive integer N, (Z^N, ω_N);
(iii) for any positive integer N, (Z^N, δ_N);
(iv) for any positive real number ϕ, (F_ϕ, β_ϕ).

3.11. Let i be an isomorphism from a digital space (V, π) to a digital space (V', π').
(i) Prove that, for any spel-adjacency ρ in (V, π), if the binary relation ρ' on V' is defined by

$$(i(c), i(d)) \in \rho' \iff (c, d) \in \rho, \tag{3.5.5}$$

then ρ' is a spel-adjacency in (V', π').
(ii) Prove that, for any surface S in (V, π) and for any spel-adjacencies κ and λ in (V, π), S is a $\kappa\lambda$-Jordan surface if, and only if, the S' of (3.4.2) is a $\kappa'\lambda'$-Jordan surface in (V', π').

3.12. For $c \in Z^2$, let $i(c) = (c_1 + c_2, c_2)$. Prove that i is an isomorphism from (Z^2, γ) to (Z^2, β).

3.13. Find an isomorphism from (Z^2, γ) to (H, ε) and prove that it has the properties which make it an isomorphism.

3.14. Prove that, for all $N \geq 2$, there are $2^{\frac{N(N-1)}{2}}$ N-dimensional sign functions s and, for each one of them, every spel has $N(N+1)$ spels β_s-adjacent to it.

3.15. Let $\bar{s}$ be the three-dimensional sign function whose value is always 1. Prove that the digital spaces $(Z^3, \beta_{\bar{s}})$ and (F, β_1) are isomorphic.

3.16. Prove in full detail that $(Z^3, \varepsilon_{(1,-1,-1)})$ is isomorphic to the bcc grid with face-adjacency.

3.17. Prove that the following are translation-invariant digital spaces:
(i) for any positive real number ϕ, (F_ϕ, β_ϕ);
(ii) for any positive integer N and for any N-dimensional direction vector e, (Z^N, ε_e).

4
Topological Digital Spaces

"Now, as I watched, the vague patch began to assemble itself, in slow dissolvings from one shape to another and still another."

R. Bradbury, *The Illustrated Man*, Epilogue.

4.1. What Is a Topology?

This chapter is for the mathematically minded. Others may skip it without any danger of losing the thread of our presentation. The content of the chapter justifies for mathematicians the approach that we have decided to take and discusses its relationship to some other possible mathematical approaches. So if you are happy with the way we are doing things and you do not particularly desire to learn more about mathematics than what is absolutely necessary to know to understand the rest of the book, just go right ahead to the next chapter.

"Topology" is a field of mathematics. Its subject matter is the study of properties of "topological spaces" the concept of which, according to Munkres [31], "grew out of the study of the real line and euclidean space and the study of continuous functions on these spaces." Topology is rich with powerful theorems. Therefore it appears desirable to investigate which of its concepts and results are applicable to digital spaces.

One of the standard concepts of topology is "connectedness." We have already introduced a notion of "connectedness" in digital spaces. In what follows we will look into the relationship between these two identically named notions.

A *topological space* is a pair $(M, \mathcal{T})$, where M is a set and $\mathcal{T}$ is a collection of subsets of M (called a *topology* on M) with the following properties:

  (i) $\emptyset$ (the empty set) and M are in $\mathcal{T}$.
  (ii) The union of elements of any subcollection of $\mathcal{T}$ is in $\mathcal{T}$.
  (iii) The intersection of elements of any finite subcollection of $\mathcal{T}$ is in $\mathcal{T}$.

Elements of $\mathcal{T}$ are referred to as *open sets*.

For a classical example, let $M = R^3$, and let $\mathcal{T}$ consist of all those sets H whose topological interior is exactly H. (Topological interiors have been defined in Chapter 2.) That this is indeed a topological space is a consequence of Exercise 4.1, which is a standard result of real analysis. According to that result, for any fixed positive integer N, if we let $\mathcal{S}$ consist of the subsets O of R^N which satisfy the property that for every $v \in O$ there exists

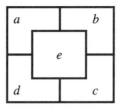

Figure 4.1.1. A simple digital space.

a positive real number r such that $B_{r,v} \subset O$ (see page 42 for the definition of $B_{r,v}$), then $\left(R^N, \mathcal{S}\right)$ is a topological space. We will refer to $\mathcal{S}$ as the *standard topology* on R^N.

Nearer to our subject matter, consider the example represented in Figure 4.1.1. There $M = \{a, b, c, d, e\}$, and a possible definition of the open sets is that $\mathcal{T}$ contains $\emptyset$, $\{a\}$, $\{c\}$, $\{a, c\}$, $\{a, b, c\}$, $\{a, c, d\}$, $\{a, b, c, d\}$, and M. It is easy to check that $\mathcal{T}$ is indeed a topology on M, according to the definition given above. (In fact, there is really no need to refer to Figure 4.1.1. The formal definitions of M and $\mathcal{T}$ in terms of the letters a, b, c, d, and e ensure that $(M, \mathcal{T})$ is a topological space. However, later on we wish to indicate that certain digital spaces which have a geometrical interpretation are in fact topological spaces. This is where the reference to Figure 4.1.1 becomes relevant.)

To further aid the reader in understanding the nature of a topological space, we give a theorem which provides us with a methodology for obtaining new topological spaces from existing ones. After the proof of the theorem, we show how it can be used to provide a particular topology on the square grid Z^2 based on the standard topology on R^2.

Before stating and proving the theorem, we give a simple example demonstrating it. Let $(M, \mathcal{T})$ be the topological space as defined above using Figure 4.1.1, and let $L = \{T, B, N\}$. In the statement of the theorem below, there is a mapping C which associates a subset of M with every element of L. If we define this mapping by $C_T = \{a, b\}$, $C_B = \{d, c\}$, and $C_N = \{e\}$, then the theorem implies (the easily checked fact) that if $\mathcal{K} = \{\emptyset, \{T, B\}, L\}$, then $(L, \mathcal{K})$ is a topological space.

Theorem 4.1.1. *Let $(M, \mathcal{T})$ be a topological space, L be an arbitrary set, and for every $c \in L$, let C_c be a subset of M. Define a collection $\mathcal{K}$ of subsets of L by*

$$K \in \mathcal{K} \Leftrightarrow \left(\bigcup_{c \in K} C_c \right) \in \mathcal{T}. \tag{4.1.1}$$

If for every $v \in M$ there is one and only one $c \in L$ such that $v \in C_c$, then $(L, \mathcal{K})$ is a topological space.

Proof. If $K = \emptyset$, then $\cup_{c \in K} C_c$ is also the empty set, and so it is an open set in the topology $\mathcal{T}$. This shows that $\emptyset \in \mathcal{K}$. Since for every $v \in M$ there is a $c \in L$ such that $v \in C_c$, $\cup_{c \in L} C_c = M$, and so $L \in \mathcal{K}$. This shows that $(L, \mathcal{K})$ satisfies condition (i) for being a topological space.

To show that it also satisfies condition (ii), let $\mathcal{I} \subset \mathcal{K}$, and let $U = \cup_{O \in \mathcal{I}} O$. We need to show that $U \in \mathcal{K}$. This follows since $\cup_{c \in U} C_c = \cup_{O \in \mathcal{I}} (\cup_{c \in O} C_c)$ is a union of open sets in the topology $\mathcal{T}$ and so is itself an open set in the topology $\mathcal{T}$.

Finally, to show that $(L, \mathcal{K})$ satisfies condition (iii) for being a topological space, let $\mathcal{I}$ be a finite subset of $\mathcal{K}$, and let $I = \cap_{O \in \mathcal{I}} O$. We need to prove that $\cup_{c \in I} C_c$ is an open set in the topology $\mathcal{T}$. We do this by showing that

$$\bigcup_{c \in I} C_c = \bigcap_{O \in \mathcal{I}} \left(\bigcup_{c \in O} C_c \right), \qquad (4.1.2)$$

which is sufficient because the right-hand side of this equation is a finite intersection of open sets in the topology $\mathcal{T}$. Recall that, for every $v \in M$, there is a unique $c \in L$ such that $v \in C_c$. It follows that v belongs to the right-hand side of (4.1.2) if, and only if, $v \in C_c$ for some fixed c which is in O for every $O \in \mathcal{I}$. In turn this is equivalent to saying that $v \in C_c$ for some fixed c in I, i.e., that v belongs to the left-hand side of (4.1.2). $\square$

As an example, we can use this theorem to define a topology on the square grid Z^2, inspired by Figure 2.2.5. With every grid point $c \in Z^2$ we associate a subset C_c of R^2, defined as follows. If c_1 and c_2 are both even, then C_c is the *open square* (see Exercise 4.2)

$$\left\{ v \in R^2 \,|\, c_1 - 1 < v_1 < c_1 + 1 \text{ and } c_2 - 1 < v_2 < c_2 + 1 \right\}. \qquad (4.1.3)$$

If c_1 is even and c_2 is odd, then C_c is the (open horizontal) line segment

$$\left\{ v \in R^2 \,|\, c_1 - 1 < v_1 < c_1 + 1 \text{ and } v_2 = c_2 \right\}. \qquad (4.1.4)$$

If c_1 is odd and c_2 is even, then C_c is the (open vertical) line segment

$$\left\{ v \in R^2 \,|\, v_1 = c_1 \text{ and } c_2 - 1 < v_2 < c_2 + 1 \right\}. \qquad (4.1.5)$$

Finally, if c_1 and c_2 are both odd, then $C_c = \{c\}$. We define the collection $\mathcal{K}$ of subsets of Z^2 by

$$K \in \mathcal{K} \Leftrightarrow \left(\bigcup_{c \in K} C_c \text{ is an open set in the standard topology on } R^2 \right). \qquad (4.1.6)$$

It is a consequence of Theorem 4.1.1 that $(Z^2, \mathcal{K})$ is a topological space.

We point out that essential use is made in the proof of the theorem of the fact that there is "only one $c \in L$ such that $v \in C_c$." If we dropped the phrase "only one" in its statement, then Theorem 4.1.1 would no longer be true; see Exercise 4.3.

Now consider an arbitrary topological space $(M, \mathcal{T})$. A subset A of M is said to be *connected* (when confusion may arise due to multiple notions of connectedness, we will use one of the expressions $\mathcal{T}$-connected or *topologically connected*), if there do **not** exist open sets O and Q such that

(i) $O \cap A \neq \emptyset$ and $Q \cap A \neq \emptyset$,
(ii) $O \cap Q \cap A = \emptyset$, and
(iii) $A \subset O \cup Q$.

This can be interpreted by saying that A is connected if it cannot be broken up into two nonempty nonintersecting parts (namely, $O \cap A$ and $Q \cap A$) using open sets. For example, now we can prove the claim made in Chapter 2 that the topological interior of the pair of sugar cubes shown in Figure 1.7.2 is not topologically connected. This is so since the topological interior of the pair of cubes is the union of the open sets which are the topological interiors of the individual sugar cubes.

To apply this well-studied notion of connectedness to the already introduced concept of connectedness in digital spaces, we need to tie them together. We say that a digital space (V, π) is a *topological digital space* if there exists a topology $\mathcal{T}$ on V, such that for any nonempty set A of spels, A is π-connected if, and only if, it is $\mathcal{T}$-connected.

As an example, consider the digital space with $V = \{a, b, c, d, e\}$ and π containing those ordered pairs of spels for which corresponding regions in Figure 4.1.1 have an edge in common. This digital space is topological with the topology $\mathcal{T}$ defined above in the discussion of the figure. To see this, first we note that the only nonempty subsets of V which are not π-connected are $\{a, c\}$ and $\{b, d\}$. That the first of these is not $\mathcal{T}$-connected can be seen by considering the two open sets $\{a\}$ and $\{c\}$. That the second one is also not $\mathcal{T}$-connected can be seen by considering the two open sets $\{a, b, c\}$ and $\{a, c, d\}$. All other nonempty subsets of V are $\mathcal{T}$-connected. For example, $\{a, b\}$ is connected since all the open sets which contain b also contain a. (If there were an O and Q satisfying (i)-(iii) above, then by (iii) one of O or Q must contain b and therefore also a. By (i), the other contains at least one of them and so (ii) would be violated.) In fact, the general results of the next section will yield immediately that this (V, π) is a topological digital space.

4.2. Some Topological Digital Spaces

For any set M, for any collection $\mathcal{T}$ of subsets of M (not necessarily a topology on M) and for any $c \in M$, we define the *smallest neighborhood* of c (with respect to $\mathcal{T}$) as

$$S_{\mathcal{T}}(c) = \bigcap_{c \in O \in \mathcal{T}} O, \qquad (4.2.1)$$

i.e., as the intersection of all sets in $\mathcal{T}$ containing c. Clearly, $c \in S_{\mathcal{T}}(c)$. We define the *induced adjacency* $\rho_{\mathcal{T}}$ on M by

$$(c, d) \in \rho_{\mathcal{T}} \Leftrightarrow c \neq d \text{ and either } c \in S_{\mathcal{T}}(d) \text{ or } d \in S_{\mathcal{T}}(c). \qquad (4.2.2)$$

Clearly, $\rho_{\mathcal{T}}$ is an adjacency (a symmetric binary relation) on M. Following [23], we call $(M, \mathcal{T})$ an *Alexandroff topological space* if, for every $c \in M$, $S_{\mathcal{T}}(c)$ is an open set. Now we illustrate these notions on the topological spaces introduced in the previous section.

First consider $(R^N, \mathcal{S})$ with $\mathcal{S}$ the standard topology on R^N. For any point $v \in R^N$, $S_{\mathcal{S}}(v) = \{v\}$, which is clearly not an open set. It follows that $(R^N, \mathcal{S})$ is not an Alexandroff topological space. Also, we see that in this case the induced adjacency is empty, i.e., for all pairs (v, w) of elements from R^N, $(v, w) \notin \rho_{\mathcal{S}}$. So the notions introduced in the previous paragraph are quite useless for euclidean spaces with the standard topology. However, as we will see, there are other spaces of interest to us for which these notions turn out to be useful.

As a baby example, consider the topological space $(M, \mathcal{T})$ associated with Figure 4.1.1. Here $S_\mathcal{T}(a) = \{a\}$, $S_\mathcal{T}(b) = \{a, b, c\}$, $S_\mathcal{T}(c) = \{c\}$, $S_\mathcal{T}(d) = \{a, c, d\}$, and $S_\mathcal{T}(e) = M$. These are all open sets, and so $(M, \mathcal{T})$ is an Alexandroff topological space. The induced adjacency is exactly the adjacency associated with Figure 4.1.1: two things are $\rho_\mathcal{T}$-adjacent if, and only if, the corresponding regions have an edge in common.

As a much more interesting example, consider the topological space $(Z^2, \mathcal{K})$ defined by (4.1.3)-(4.1.6). Let $c \in Z^2$. If c_1 and c_2 are both even, then $S_\mathcal{K}(c) = \{c\}$ (because $\{c\}$ is open by Exercise 4.2 and is clearly a subset of any set containing c). If c_1 is even and c_2 is odd, then

$$S_\mathcal{K}(c) = \left\{ d \in Z^2 \,|\, d_1 = c_1, \, c_2 - 1 \le d_2 \le c_2 + 1 \right\} \tag{4.2.3}$$

(because this set is open, as can be easily proved based on (4.1.3), (4.1.4) and (4.1.6), and has to be a subset of any open set containing c, since for all positive r the ball $B_{r,c}$ intersects both of the two open squares above and below the horizontal line segment defined by (4.1.4)). Similarly, if c_1 is odd and c_2 is even, then

$$S_\mathcal{K}(c) = \left\{ d \in Z^2 \,|\, c_1 - 1 \le d_1 \le c_1 + 1, \, d_2 = c_2 \right\} \tag{4.2.4}$$

is an open set, and if c_1 and c_2 are both odd, then

$$S_\mathcal{K}(c) = \left\{ d \in Z^2 \,|\, c_1 - 1 \le d_1 \le c_1 + 1, \, c_2 - 1 \le d_2 \le c_2 + 1 \right\} \tag{4.2.5}$$

is an open set. Therefore it follows that $(Z^2, \mathcal{K})$ is an Alexandroff topological space. Furthermore, by looking at Figure 2.2.5, we see that the induced adjacency $\rho_\mathcal{K}$ is exactly the Khalimsky adjacency on Z^2. (Surprised?)

Lemma 4.2.1. *For any Alexandroff topological space $(M, \mathcal{T})$ and for any subset A of M, A is $\mathcal{T}$-connected if, and only if, A is $\rho_\mathcal{T}$-connected.*

Proof. First suppose that A is $\rho_\mathcal{T}$-connected. Now we show that if there exist open sets O and Q such that $O \cap A \ne \emptyset$, $Q \cap A \ne \emptyset$, and $A \subset O \cup Q$, then $O \cap Q \cap A \ne \emptyset$. (This implies, by definition, that A is $\mathcal{T}$-connected.) Let $c \in O \cap A$ and $d \in Q \cap A$. By the $\rho_\mathcal{T}$-connectedness of A, there exists a $\rho_\mathcal{T}$-path $\langle c^{(0)}, \cdots, c^{(K)} \rangle$ in A from c to d. Since we have assumed that $A \subset O \cup Q$, there must be a k, $1 \le k \le K$, such that $c^{(k-1)} \in O$ and $c^{(k)} \in Q$. Since $(c^{(k-1)}, c^{(k)}) \in \rho_\mathcal{T}$, either $c^{(k-1)} \in S_\mathcal{T}(c^{(k)})$ or $c^{(k)} \in S_\mathcal{T}(c^{(k-1)})$. We complete this first part of the proof assuming that the latter is the case; the proof of the alternative is strictly analogous. Since O is an open set which contains $c^{(k-1)}$, it follows from the definition of the smallest neighborhood that $S_\mathcal{T}(c^{(k-1)}) \subset O$ and so $c^{(k)} \in O \cap Q \cap A$.

Conversely, suppose that A is not $\rho_\mathcal{T}$-connected, and let a and b be two elements of A for which there is no $\rho_\mathcal{T}$-path in A connecting them. Let C be the set of all elements in A which are $\rho_\mathcal{T}$-connected in A to a, and let D be the set of all the other elements of A. Further, let $O = \cup_{c \in C} S_\mathcal{T}(c)$ and $Q = \cup_{d \in D} S_\mathcal{T}(d)$. Both O and Q are open since they are unions of open sets. Neither $O \cap A$ nor $Q \cap A$ are empty (they contain a and b, respectively). To show that $O \cap Q \cap A = \emptyset$, let us assume that, on the contrary, there is an $e \in O \cap Q \cap A$. Since $A = C \cup D$, either $e \in C$ or $e \in D$. We assume the former; the proof in the latter case is strictly analogous. Since $e \in Q$, there is a $d \in D$ such that $e \in S_\mathcal{T}(d)$. Furthermore, $e \ne d$ (since $C \cap D = \emptyset$) and so, according to (4.2.2), $(e, d) \in \rho_\mathcal{T}$. By definition of C, there is a ρ_T-path in A connecting a to e and, by what we have just shown, this can be extended to a ρ_T-path in A connecting a to d. This contradicts the fact that $d \in D$ (which implies that $d \notin C$). Finally, since $C \subset O$ and $D \subset Q$, $A \subset O \cup Q$. This proves that A is not T-connected. $\square$

Based on the discussion just before the statement of this lemma, it is quite obvious that for the topological space (M, T) associated with Figure 4.1.1, M is ρ_T-connected and for the topological space (Z^2, K) defined by (4.1.3)-(4.1.6), Z^2 is ρ_K-connected. Lemma 4.2.1 implies that in the former case M is T-connected and that in the latter case Z^2 is K-connected.

Theorem 4.2.2. *For any Alexandroff topological space (V, T) with a nonempty V, (V, ρ_T) is a topological digital space if, and only if, V is T-connected.*

Proof. By definition, (V, ρ_T) is a digital space if, and only if, V is ρ_T-connected. By Lemma 4.2.1, this can happen if, and only if, V is T-connected. The same lemma implies that if this condition is satisfied, then (V, ρ_T) is necessarily a topological digital space. $\square$

Now the paragraph just before the statement of this theorem is seen to imply that the two digital spaces associated with Figures 4.1.1 and 2.2.5, respectively, are both topological. To exploit Theorem 4.2.2 further to show that certain digital spaces are topological, we narrow our approach to a particular way of specifying collections of subsets.

Lemma 4.2.3. *For any set M and for any $c \in M$, let $B(c)$ be a subset of M which contains c. Let T be defined as the collection of those subsets O of M for which the following is true: for every $c \in O$, $B(c) \subset O$. Then, (M, T) is an Alexandroff topological space, and for any $c \in M$, $B(c) \subset S_T(c)$.*

Proof. Let T be defined as in the statement of the lemma. The empty set is in T, vacuously. For any $c \in M$, $B(c) \subset M$, which implies that $M \in T$. Now consider a c which belongs to a union of open sets. Then it belongs to at least to one of the open sets forming the union, say to O, and so $B(c) \subset O$. Since O is a subset of the union of open sets to which c belongs, this shows that the union itself is open. Finally, consider the case when c belongs to the intersection of (a not necessarily finite) subcollection of T. Then $B(c) \subset O$ for every open set O in the subcollection, which implies that $B(c)$ is a subset of the intersection of these open sets. This proves that the intersection is open and, in view of (4.2.1), that $B(c)$ is a subset of the open set $S_T(c)$. $\square$

As an illustration, let $M = \{a, b, c, d, e\}$, and let $B(a) = \{a\}$, $B(b) = \{a, b, c\}$, $B(c) = \{c\}$, $B(d) = \{a, c, d\}$, and $B(e) = \{b, d, e\}$. It is easy to see that the method specified in the previous lemma for defining open sets results in $T = \{\emptyset, \{a\}, \{c\}, \{a, c\}, \{a, b, c\}, \{a, c, d\}, \{a, b, c, d\}, M\}$, which is exactly the topology associated with Figure 4.1.1, and so we have shown (yet again) that this particular (M, T) is an Alexandroff topological space. Note that $B(e)$ is **not** an open set since any open set that contains b must contain the whole of $B(b)$. In fact $S_T(e) = M$, which shows that the containment at the end of the statement of Lemma 4.2.3 cannot be replaced by an equality.

To use Theorem 4.2.2 for proving that some more interesting digital spaces are topological, we introduce a further concept. For any set M, the mapping B from M into subsets of M is said to be a *basis mapping* on M if for all $c \in M$, $c \in B(c)$ and $B(c) = \bigcup_{d \in B(c)} B_d$.

Theorem 4.2.4. *A digital space* (V, π) *is topological, provided that there exists a basis mapping B on V such that, for all c and d in V,*

$$(c, d) \in \pi \Leftrightarrow c \neq d \text{ and either } c \in B(d) \text{ or } d \in B(c). \tag{4.2.6}$$

Proof. Using B, define $\mathcal{T}$ as in the statement of Lemma 4.2.3. According to that lemma, $(V, \mathcal{T})$ is an Alexandroff topological space and, for any $c \in V$, $B(c) \subset S_{\mathcal{T}}(c)$. Since B is a basis mapping, for every $d \in B(c)$ we have that $B(d) \subset B(c)$, which by definition implies that $B(c)$ is open, and hence $B(c) = S_{\mathcal{T}}(c)$. Comparing (4.2.2) with (4.2.6), we see that this implies that $\rho_{\mathcal{T}} = \pi$. By definition of a digital space, V is π-connected, and so by Lemma 4.2.1, V is $\mathcal{T}$-connected. Knowing this, Theorem 4.2.2 yields that (V, π) is a topological digital space. $\square$

Theorem 4.2.5. *For every positive integer N, the digital space (Z^N, ω_N) is topological.*

Proof. For every $c \in Z^N$, define the set

$$B(c) = \begin{cases} \{c\}, & \text{if } \sum_{n=1}^{N} c_n \text{ is even,} \\ \{c\} \cup \{d \mid (c, d) \in \omega_N\}, & \text{otherwise.} \end{cases} \tag{4.2.7}$$

By observing that if two spels are ω_N-adjacent then the sum of the components of one of them must be even and of the other of them must be odd, we see that the B of the equation above is a basis mapping on Z^N and (4.2.6) is satisfied with ω_N in place of π. Hence the result follows from Theorem 4.2.4. $\square$

Another family of topological digital spaces comes from the Khalimsky adjacencies whose two-dimensional manifestation was discussed at the end of Chapter 2. Here we will define a Khalimsky adjacency on Z^N for an arbitrary $N \geq 1$, but first introduce a preliminary concept which will also be found useful in other places in this book.

For any positive integer N, a binary relation ρ on Z^N is said to be *local* if $\rho \subset \alpha_N$; see page 8. The geometrical interpretation of local is obvious: if ρ is a local binary relation on Z^N, then $(c, d) \in \rho$ implies that the intersection of the Voronoi neighborhoods of c and of d in Z^N cannot be empty; in symbols, $N_{Z^N}(c) \cap N_{Z^N}(d) \neq \emptyset$. We leave it to the reader to check that, for any N-dimensional sign function s and for any N-dimensional direction vector e, both β_s and ε_e are local and that a local binary relation on Z^N is finitary.

For any positive integer N, the *Khalimsky adjacency* κ_N is the local adjacency on Z^N characterized by $(c, d) \in \kappa_N$ if, and only if, either c_n is odd for all such n that $|c_n - d_n| = 1$ or c_n is even for all such n that $|c_n - d_n| = 1$. (Note that the locality implies that $c \neq d$.) We leave it to the reader to prove that in the lower dimensional cases this definition corresponds to the concept introduced at the end of Chapter 2.

Theorem 4.2.6. *For every positive integer N, κ_N is a spel-adjacency in (Z^N, ω_N).*

Proof. We need to prove that κ_N is symmetric and that $\omega_N \subset \kappa_N$. Symmetry follows from the fact that c_n is odd for all n such that $|c_n - d_n| = 1$ if, and only if, d_n is even for all n such that $|d_n - c_n| = 1$. That $\omega_N \subset \kappa_N$ follows from the fact that two ω_N-adjacent elements of Z^N differ at exactly one coordinate and the difference at that coordinate is 1. $\square$

Theorem 4.2.7.   *For every nonnegative integer N, the digital space $\left(Z^N, \kappa_N\right)$ is topological.*

Proof. For every $c \in Z^N$, define the set

$$B(c) = \{d \mid (d_n = c_n, \text{ if } c_n \text{ is even}) \text{ and } (|d_n - c_n| \leq 1, \text{ if } c_n \text{ is odd})\}. \qquad (4.2.8)$$

It is worthwhile here to diverge from the formal proof for an intuitive understanding of these sets. Consider the case $N = 2$ and look at Figure 2.2.5. A vector c whose components are all even is interpreted in that figure as a pixel in $2Z$. For such a vector $B(c)$ consists of only of this pixel. A vector c which has exactly one even component is interpreted as an edge of a pixel in $2Z$. For such a vector $B(c)$ consists of the edge itself and the two pixels which share that edge. A vector c whose components are all odd is interpreted as a vertex of a pixel in $2Z$. For such a vector $B(c)$ consists of the vertex itself, the four edges which meet at that vertex, and the four pixels which share that vertex. Yet an alternative way of looking at this is to observe that, in the two-dimensional case, the $B(c)$ are exactly the $S_\kappa(c)$ defined just before Lemma 4.2.1

Now we show that B of (4.2.8) is a basis mapping on Z^N. Consider any spels c, d, and e such that $d \in B_c$ and $e \in B_d$. We need to show that $e \in B_c$. If c_n is even, then d_n is even and so $e_n = d_n = c_n$. If c_n is odd and $d_n = c_n$, then $|e_n - c_n| = |e_n - d_n| \leq 1$. If c_n is odd and $|d_n - c_n| = 1$, then d_n is even and so $e_n = d_n$, which implies that $|e_n - c_n| = 1$.

Next we show that if ρ is defined by (4.2.6), with B as in (4.2.8), then $\rho = \kappa_N$. First suppose that $(c, d) \in \rho$. Then $c \neq d$, and either $c \in B(d)$ or $d \in B(c)$. In the second case, according to (4.2.8), c_n and d_n differ at most by 1 and, if they differ, c_n has to be odd. In the first case, according to (4.2.8), c_n and d_n differ at most by 1 and, if they differ, d_n has to be odd and hence c_n has to be even. In either case, $(c, d) \in \kappa_N$. Now suppose that $(c, d) \in \kappa_N$. We note that, since κ_N is local, if $|c_n - d_n| \neq 1$, then $|c_n - d_n| = 0$. Observing (4.2.8), we see that if c_n is odd for all n such that $|c_n - d_n| = 1$, then $d \in B(c)$ and if c_n is even for all n such that $|c_n - d_n| = 1$, then $c \in B(d)$. In either case, $(c, d) \in \rho$.

Now we have shown that the B of (4.2.8) is a basis mapping on Z^N and that (4.2.6) is satisfied with κ_N in place of π. Hence the result follows from Theorem 4.2.4. $\square$

4.3. Many Digital Spaces Are Not Topological

In the previous section we described two infinite families of topological digital spaces. Unfortunately, those families just about exhaust all the interesting examples of digital spaces which are topological. We can say this because we have the following powerful theorem to show that a very large class of digital spaces is not topological. A special case of this theorem has been stated and proved in [28]. The proof given there is essentially the same as the proof for the general case.

Theorem 4.3.1.   *If the digital space (V, π) is such that V contains distinct elements $c^1, c^2, c^3, d^1, d^2, d^3$ such that*

$$\left(c^i, c^j\right) \in \pi, \text{ if } i \neq j \qquad (4.3.1)$$

and

$$\left(c^i, d^i\right) \in \pi, \ \ but \ \left(c^i, d^j\right) \notin \pi \ if \ i \neq j \, , \tag{4.3.2}$$

then (V, π) is not topological.

Proof. Let $\mathcal{T}$ be any topology on V such that every $\mathcal{T}$-connected set of spels is π-connected. By making various assumptions regarding (V, π), which between them exhaust all possible cases, we will produce π-connected sets of spels which are not $\mathcal{T}$-connected. We use C to denote $\{c^1, c^2, c^3\}$.

First assume that, for $1 \leq i \leq 3$, there is an open set O_i not containing c^i such that $O_i \cap C \neq \emptyset$. Without loss of generality, we may also assume that $c^2 \in O_3$. If $c^3 \in O_2$, then the π-connected set $\{c^2, c^3\}$ is not $\mathcal{T}$-connected (as can be seen by considering the open sets O_2 and O_3). If $c^3 \notin O_2$, then $O_2 \cap C = \{c^1\}$. It follows that the π-connected set $(O_1 \cap C) \cup \{c^1\}$ is not $\mathcal{T}$-connected (as can be seen by considering the open sets O_1 and O_2).

To deal with the alternative case, we may assume (without loss of generality) that whenever an open set has a nonempty intersection with C, then it contains c^3. Since neither of the sets $\{c^1, d^2\}$ and $\{c^2, d^1\}$ is π-connected, it follows from the assumed nature of $\mathcal{T}$ that they are also not $\mathcal{T}$-connected. In particular, this implies that there are open sets U_1 and U_2, such that $c^1 \in U_1$, $d^2 \notin U_1$, $c^2 \in U_2$, and $d^1 \notin U_2$. If neither $C \subset U_1$ nor $C \subset U_2$, then the π-connected set $\{c^1, c^2\}$ is not $\mathcal{T}$-connected (as can be seen by considering the open sets U_1 and U_2). Assuming that $C \subset U_1$ (the assumption $C \subset U_2$ can be treated similarly), we observe that there is an open set W such that $c^3 \notin W$ and $d^2 \in W$ (because $\{c^3, d^2\}$ is not π-connected). Our assumption on the nature of c^3 implies that $W \cap C = \emptyset$. From this follows that the π-connected set $\{c^2, d^2\}$ is not $\mathcal{T}$-connected (as can be seen by considering the open sets U_1 and W). $\square$

We demonstrate the power of this theorem by giving two examples of digital spaces which can be seen as not topological as a result of it. The first example is the one discussed in [28]. It can be seen from Figure 4.3.1 that it is a consequence of Theorem 4.3.1 that the space $\left(Z^2, \alpha_2\right)$ is not topological. More importantly from the point of view of our general approach, the digital space associated with the hexagonal tessellation of the plane is also not topological, as can be seen from Figure 4.3.2.

Clearly, there are numerous other spaces which are not topological as a consequence of this theorem. We leave it to the reader to prove that, for any N greater than 1, the digital

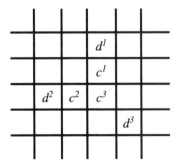

Figure 4.3.1. Demonstration, based on Theorem 4.3.1, that the space $\left(Z^2, \alpha_2\right)$ is not topological.

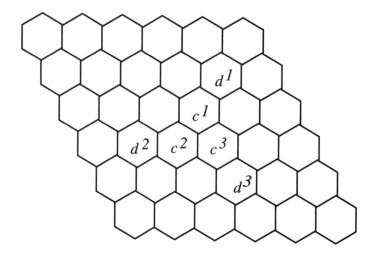

Figure 4.3.2. Demonstration, based on Theorem 4.3.1, that the digital space based on the hexagonal tessellation of the plane and edge-adjacency is not topological.

spaces $\left(Z^N, \alpha_N\right)$, $\left(Z^N, \beta_s\right)$ for an arbitrary N-dimensional sign function s, and $\left(Z^N, \varepsilon_e\right)$ for an arbitrary N-dimensional direction vector e are not topological. Our conclusion from this is that we cannot, without a serious loss of its general applicability, embed our theory into one in which the connectedness of a set of spels is defined as topological connectedness due to some topology (in the classical sense). For this reason, we cannot expect our theory of digital spaces to be derivable from established results of classical topology.

4.4. Connectedness of Topological Interiors

In this section we return to an important issue raised near the beginning of Chapter 2. First we reformulate the issue in the more precise mathematical terminology that we have since developed. Let (V, π) be a digital space for which V is a Delone grid in R^N. What conditions on π ensure that the following desirable property holds: for every π-connected set of spels O, the topological interior of the union of the Voronoi neighborhoods of O is topologically connected?

Previously we have dealt with this issue only when $N = 3$. However, much that we have to say is just as easily stated for an arbitrary positive N, and it is not even necessary to restrict our discussion to Voronoi neighborhoods. Therefore we proceed to treat the problem in this more general setting. First we need to generalize some previously given definitions. The *topological interior* of a subset H of R^N is the set of all $v \in H$ for which there exists a positive real number r such that $B_{r,v} \subset H$. Let S consist of all those sets H whose topological interior is exactly H. We have left it to the reader to prove, as Exercise 4.1, that S is a topology on R^N; we refer to it as the standard topology. In this section, when we refer to a topologically connected set, we will mean by it an S-connected subset of R^N.

Our general approach uses certain types of "star-convex sets" (see, e.g., [31]). At the end of the section we will point out the possibility of an even more general setting, but for now we concentrate on making precise the types of subsets of R^N which will be the subject of our discourse in most of this section.

The *line segment* joining v to w in R^N is defined as the set $\{v + t(w - v) \mid 0 \le t \le 1\}$. (As is easily seen, this is the same set as the line segment joining w to v. Nevertheless, it is useful for us to distinguish between the beginning and the end of a line segment.) A subset H of R^N is said to be a *strictly star-convex set* if there is an element v in H such that, for every w in H, all elements of the line segment joining v to w, with the possible exception of w, are in the topological interior of H. (The set of such v in H form the *core* of H.) An illustration of such sets in R^2 is provided in Figure 4.4.1. Further on we discuss other examples of strictly star-convex sets (Voronoi neighborhoods and convex polyhedra), but first we wish to state and prove the result which makes them interesting.

We call a subset H of R^N *linearly path connected* if, for every v and w in H, there exists a sequence $\langle v = v^{(0)}, \cdots, v^{(L)} = w \rangle$ of elements of R^N such that for $1 \le l \le L$ the line segment joining $v^{(l-1)}$ to $v^{(l)}$ is a subset of H. It is a standard result of topology (see, e.g., [31], page 155) that a linearly path connected subset of R^N is a topologically connected set.

Theorem 4.4.1. *For any positive integer N, let M be a collection of strictly star-convex sets in R^N, and define the binary relation ρ on M by*

$$(C, D) \in \rho \Leftrightarrow C \ne D \text{ and } C \cap D \text{ intersects the topological interior of } C \cup D. \quad (4.4.1)$$

If A is a ρ-connected subset of M, then the topological interior of $\bigcup_{C \in A} C$ is a topologically connected set.

Proof. By the paragraph just before the statement of the theorem, it is sufficient to prove that, for every v and w in the topological interior of $\cup_{C \in A} C$, there exists a sequence $\langle v = v^{(0)}, \cdots, v^{(L)} = w \rangle$ of elements of R^N such that for $1 \le l \le L$ the line segment joining $v^{(l-1)}$ to $v^{(l)}$ is a subset of the topological interior of $\cup_{C \in A} C$. Let us denote by C an element of A which contains v and by D an element of A which contains w. Since A is a ρ-connected subset of M, there is a ρ-path $\langle C^{(0)}, \cdots, C^{(K)} \rangle$ in A from C to D.

The desired sequence is defined as follows. Let $L = 2K + 2$. For $0 \le k \le K$, let $v^{(2k+1)}$ be an element of the core of $C^{(k)}$ and, for $1 \le k \le K$, let $v^{(2k)}$ be a common element of $C^{(k-1)} \cap C^{(k)}$ and the topological interior of $C^{(k-1)} \cup C^{(k)}$. It is easy to check that each of the resulting line segments is a subset of the topological interior of $\cup_{C \in A} C$. For example, if $1 \le k \le K$, then $v^{(2k-1)}$ is an element of the core of the strictly star-convex set $C^{(k-1)}$ and $v^{(2k)} \in C^{(k-1)}$. Hence, all elements of the line segment joining $v^{(2k-1)}$ to $v^{(2k)}$, with the possible exception of $v^{(2k)}$, are in the topological interior of $C^{(k-1)}$ and, consequently (see Exercise 4.12), in the topological interior of $\cup_{C \in A} C$. On the other hand, $v^{(2k)}$ is in the topological interior of $C^{(k-1)} \cup C^{(k)}$ and, consequently (see Exercise 4.12), in the topological interior of $\cup_{C \in A} C$. $\square$

Although we wanted to make sure that our theory is general enough to deal with tessellations such as illustrated in Figure 4.4.1, typically we restrict our attention to a more special class of strictly star-convex sets. A subset H of R^N is said to be a *convex set* if, for every v and w in H, all elements of the line segment joining v to w are in H. From the point of view of our approach, convexity is not a serious restriction, because of the following result.

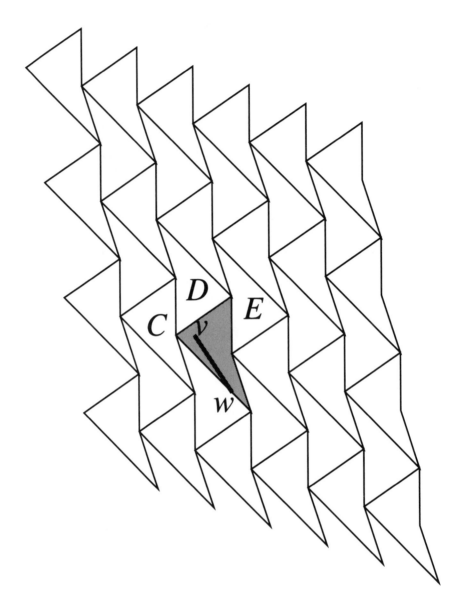

Figure 4.4.1. Some strictly star-convex sets in R^2; they contain the interiors and the edges of the indicated polygons. One of them is shaded, and for this one the nature of its strictly star-convexity is illustrated for a single element v of its core. (Similar sets are discussed on page 148 of [40].) The sets C and D are ρ-adjacent according to (4.4.1), but the sets D and E are not (since the single common point in their intersection is not in the topological interior of their union).

Lemma 4.4.2.   *For any grid G in R^N and for any $g \in G$, $N_G(g)$ is convex.*

Proof. Suppose that v and w are in $N_G(g)$. We need to prove that, for any $h \in G$ and for $0 \le t \le 1$,

$$\|v + t(w - v) - g\| \le \|v + t(w - v) - h\|. \qquad (4.4.2)$$

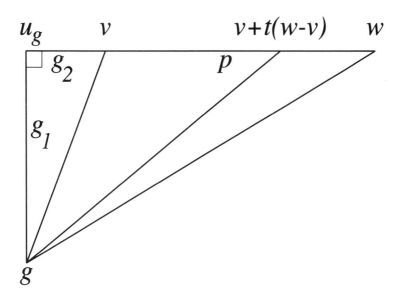

Figure 4.4.2. Illustration of the convexity of Voronoi neighborhoods.

Now let u_g be the orthogonal projection of g onto the line containing v and w (i.e., u_g has the form $v + s(w - v)$ for some real number s and $(g - u_g) \cdot (w - v) = 0$; see Figure 4.4.2. Let $g_1 = \|g - u_g\|$, $g_2 = \|v - u_g\|$ and $p = \|w - v\|$. Then,

$$\|v + t(w - v) - g\|^2 = (1 - t)\left(g_1^2 + g_2^2\right) + t\left(g_1^2 + g_2^2 + 2pg_2\right) + t^2p^2 . \qquad (4.4.3)$$

Now if we define u_h, h_1, and h_2 analogously, we get that

$$\|v + t(w - v) - h\|^2 = (1 - t)\left(h_1^2 + h_2^2\right) + t\left(h_1^2 + h_2^2 + 2ph_2\right) + t^2p^2 . \qquad (4.4.4)$$

Since (4.4.2) holds if $t = 0$ or if $t = 1$, we get that $g_1^2 + g_2^2 \leq h_1^2 + h_2^2$ and that $g_1^2 + g_2^2 + 2pg_2 \leq h_1^2 + h_2^2 + 2ph_2$. It follows that (4.4.2) holds as long as $0 \leq t \leq 1$.
□

To tie this result to Theorem 4.4.1, now we show the relationship between convexity and strictly star-convexity.

Lemma 4.4.3. *Any convex set in R^N which has a nonempty topological interior is a strictly star-convex set, and its core is its topological interior.*

Proof. Let v be any element of the topological interior of a convex set H in R^N, and let w be any other element of H. In the next paragraph we show that if $u = v + t(w - v)$, for $0 \leq t < 1$, then u is in the topological interior of H. This is sufficient to prove that H is a strictly star-convex set and that its topological interior is a subset of its core. An element of H which is not in its topological interior cannot be in its core, since the line segment from it to any element in the topological interior would not have the property used in the definition

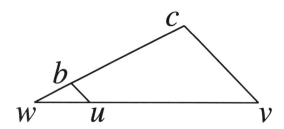

Figure 4.4.3. Location of the vectors used in the proof of Lemma 4.4.3.

of the core of a strictly star-convex set. Consequently, the material of the next paragraph is all we need to complete the proof.

Since v is an element of the topological interior of H, there exists a positive real number r such that $B_{r,v} \subset H$. We complete the proof by showing that $B_{r(1-t),u} \subset H$ (recall that $0 \le t < 1$). Consider an arbitrary element b of $B_{r(1-t),u}$. Such an element has the form $b = u + si$, where $i \in R^N$, $\|i\| = 1$ and $0 \le s \le r(1-t)$. Let $c = v + \frac{s}{1-t}i$, see Figure 4.4.3. Clearly, c is in $B_{r,v}$ and hence in H. By convexity of H, all points of the line segment joining c to w are in H. In particular, $b \in H$ since

$$b = u + si = v + t(w - v) + si = v + \frac{s}{1-t}i + t\left(w - v - \frac{s}{1-t}i\right) = c + t(w - c). \quad (4.4.5)$$

$\square$

Theorem 4.4.4. *Let G be any Delone grid in R^N, and define the binary relation π on G by*

$$(c,d) \in \pi \Leftrightarrow c \ne d \text{ and } N_G(c) \cap N_G(d) \text{ intersects the topological interior of } N_G(c) \cup N_G(d).$$
$$(4.4.6)$$

If O is a π-connected subset of G, then the topological interior of $\bigcup_{c \in O} N_G(c)$ is a topologically connected set.

Proof. By Lemma 4.4.2, $N_G(g)$ is convex for any $g \in G$. Furthermore, the topological interior of $N_G(g)$ is not empty (in particular, it contains g; see Exercise 1.1). It follows from Lemma 4.4.3 that $N_G(g)$ is strictly star-convex. Therefore we can define $M = \{N_G(g) \mid g \in G\}$ and appeal to Theorem 4.4.1. We note that, for our M, the ρ of that theorem has the property that

$$(c,d) \in \pi \Leftrightarrow (N_G(c), N_G(d)) \in \rho. \quad (4.4.7)$$

It follows that if O is a π-connected subset of G, then $A = \{N_G(g) \mid g \in O\}$ is a ρ-connected subset of M and so, by Theorem 4.4.1, the topological interior of $\cup_{c \in O} N_G(c) = \cup_{C \in A} C$ is a topologically connected set. $\square$

A little reflection will show that the adjacency defined by (4.4.6) is a mathematically rigorous generalization of what we called edge-adjacency when $N = 2$ and face-adjacency when $N = 3$. By looking at Figures 1.4.1 and 1.6.2, we see that when $G = Z^2$ the π of (4.4.6) is ω_2 and when $G = Z^3$ the π of (4.4.6) is ω_3. In fact, we claim that, for any positive N, if $G = Z^N$, then the π of (4.4.6) is ω_N. Although this is a desirable property of these adjacencies, they fail to have another desirable property: if we try to use them as the single adjacency relation for both the foreground (sugar cubes) and background, then it cannot be guaranteed that a boundary between a component of the foreground and a component of the background will be a Jordan surface. (This was repeatedly discussed in the previous chapters.) This is where some alternative grids are clearly advantageous. Looking at Figures 2.2.3 and 2.1.5, we see that the adjacency defined by (4.4.6) is the edge-adjacency for the hexagonal grid and the face-adjacency β_ϕ for the fcc grid F_ϕ. However, these adjacencies also have the desired Jordan-type properties, as stated, for example, in Claim 2.1.1 and as will be rigorously proved in the forthcoming chapters.

Now we diverge a little bit to discuss a class of tessellations of N-dimensional euclidean spaces which is a popular area of study. This class is more general than the one provided by the Voronoi neighborhoods of Delone grids but is not as general as the one provided by strictly star-convex sets. Since this particular class of tessellations will not play a significant role in this book, we will keep our discussion at a somewhat superficial level.

With any element a of R^N and with any real number b, we associate a *half-space* which is the subset of R^N defined by $H = \{ v \mid a \cdot v \le b \}$. We call that subset of such an H which is defined by $\partial H = \{ v \mid a \cdot v = b \}$ the *bounding plane* of H. We call an intersection of finitely many half-spaces in R^N a *convex polyhedron* if it is bounded and has a nonempty topological interior. A set F is called a *facet of a convex polyhedron* P in R^N, if there exists a half-space H such that $P \subset H$ and $P \cap \partial H = F$. (Note that according to this definition, the empty set is a facet of every convex polyhedron.) A *polyhedral tessellation* $\mathcal{P}$ of R^N is a collection of convex polyhedra such that

  (i) for every $v \in R^N$, there is a $P \in \mathcal{P}$ such that $v \in P$;
  (ii) if P and Q are distinct elements of $\mathcal{P}$, then there is a facet F of P and a facet G of Q such that $P \cap Q = F \cap G$.

As an illustration, consider Figure 4.4.4. Here the intended tessellation of the plane is into square-shaped "bricks" of unit area. Every brick is labeled by a $c \in Z^2$. With each such c, we associate four half-spaces defined as follows:

$$H_c^u = \{ v \mid v_2 \le c_2 + 0.5 \} ; \tag{4.4.8}$$

$$H_c^d = \{ v \mid -v_2 \le -c_2 + 0.5 \} ; \tag{4.4.9}$$

$$H_c^r = \{ v \mid v_1 \le c_1 - (c_2 - 1)/2 \} ; \tag{4.4.10}$$

$$H_c^l = \{ v \mid -v_1 \le -c_1 + (c_2 + 1)/2 \} . \tag{4.4.11}$$

The bounding planes of these four half-spaces when $c = (2, 2)$ are indicated by the double lines in Figure 4.4.4. The arrows indicate the side of the bounding plane occupied by the half-space. The intersection of these four half-spaces is indeed the closed square labeled by $(2, 2)$. Such a square is bounded and has a nonempty topological interior. So it is a convex polyhedron according to the definition given above. Its facets are its four edges, its four vertices, and the empty set. The collection of all such squares for $c \in Z^2$ is a polyhedral

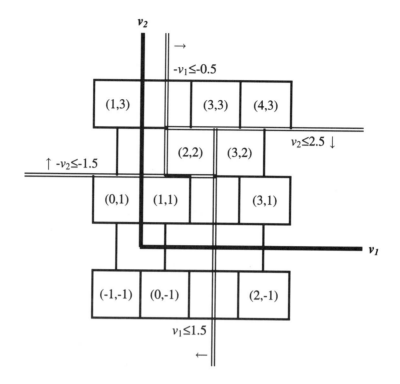

Figure 4.4.4. The polyhedral tessellation of the plane defined by (4.4.8)-(4.4.11).

tessellation of R^2, as is clear by looking at Figure 4.4.4. For example, if P is the closed square associated with $(2, 2)$ and Q is the closed square associated with $(1, 1)$, then $P \cap Q$ is indeed the intersection of the lower edge of P with the upper edge of Q, as indicated by the heavy line segment in the figure.

The last observation also shows that the collection of squares depicted in Figure 4.4.4 cannot possibly be the collection of the Voronoi neighborhoods of a Delone grid in R^2 since such a collection of Voronoi neighborhoods consists of polygons which must share a whole edge if they share a point which is not a vertex of both; see Exercise 1.5. In fact, polyhedral tessellations are a generalization of such collections of Voronoi neighborhoods, as is stated in Exercise 4.16.

We also leave it to the reader to show that every convex polyhedron in R^N in fact is a convex set in R^N. By definition, a convex polyhedron has a nonempty topological interior, and so (by Lemma 4.4.3) it is a strictly star-convex set. Hence, we can use Theorem 4.4.1 to define an adjacency relation on the convex polyhedra of a polyhedral tessellation of R^N and be sure that the topological interior of the union of a connected set of convex polyhedra will be topologically connected. For example, for the tessellation illustrated in Figure 4.4.4, the closed squares associated with $(2, 2)$ and with $(1, 1)$ are ρ-adjacent according to (4.4.1). The heavy line segment in Figure 4.4.4 indicates the set of points which are both in the intersection of the two closed squares and in the topological interior of their union. From the point of view of digital geometry, however, nothing new has been introduced by this particular example: we leave it to the reader to show that the digital space whose elements are the closed squares of Figure 4.4.4 and whose proto-adjacency is defined by (4.4.1) is isomorphic to the hexagonal tessellation of the plane with edge-adjacency, as illustrated in Figure 2.2.3.

We complete this section by discussing how its approach can be generalized. The notion of a linearly path connected set can be replaced by the notion of a path connected set [31] in which every pair of points can be connected by a "continuous curve" (rather than by just a sequence of line segments). Similarly, the notion of a strictly star-convex set can be replaced by a set in which any two points can be connected by a "continuous curve" which lies (except for its initial and final points) in the interior of the set. Alternatively, we can look at the sets as defined before and see what happens to them under a "rubber sheet" transformation. In either case, we end up with a much more general class of tessellations of euclidean spaces, but the interpretation of them using the approach of digital spaces does not get any more complicated. (This is because, using the straightforward generalization of Theorem 4.4.1, we still end up with a set and an adjacency defined on the set; all the geometrical details of the underlying continuous space become secondary compared to the questions that we ask regarding the overlying digital space.)

4.5. Exercises

4.1. For any fixed positive integer N, let $\mathcal{S}$ consist of the subsets O of R^N which satisfy the property that, for every $v \in O$, there exists a positive real number r such that $B_{r,v} \subset O$ (see page 42 for the definition of $B_{r,v}$). Prove that $(R^N, \mathcal{S})$ is a topological space.

4.2. Prove that for any $c \in Z^2$, the subset of R^2 defined by (4.1.3) is an open set in the standard topology on R^2.

4.3. Let $M = \{a, b, c\}$ and $\mathcal{T} = \{\emptyset, \{a\}, \{c\}, \{a, b\}, \{a, c\}, \{a, b, c\}\}$.
 (i) Prove that $(M, \mathcal{T})$ is a topological space.
 (ii) Let $L = \{u, v, w\}$, define $C_u = \{a\}$, $C_v = \{a, b\}$, $C_w = \{b, c\}$, and use (4.1.1) to define $\mathcal{K}$. Prove that $(L, \mathcal{K})$ is not a topological space.
 (iii) Find the subset $\mathcal{I}$ of $\mathcal{K}$ for which (4.1.2) fails to hold.

4.4. For the standard topology $\mathcal{S}$ on R^N show that, for any $v \in R^N$, $S_{\mathcal{S}}(v) = \{v\}$.

4.5. Prove that there is a topological space $(V, \mathcal{T})$ which is not Alexandroff, but for which $(V, \rho_{\mathcal{T}})$ is a topological digital space. (*Hint:* Let V be the set of nonnegative real numbers, and let each interval $[0, r) = \{v \mid 0 \le v < r\}$ be an open set.)

4.6. Let $M = \{a, b, c\}$, and let $B(a) = \{a, b\}$, $B(b) = \{b, c\}$, and $B(c) = \{a, c\}$. What is the topology on M generated on the basis of this function by the method described in the statement of Lemma 4.2.3?

4.7. For any positive integer N, show that

(i) for any N-dimensional sign function s, β_s is local;

(ii) for any N-dimensional direction vector e, ε_e is local;

(iii) if ρ is a local binary relation on Z^N, then ρ is also finitary.

4.8.

(i) Prove that the κ_2 defined on page 87 is exactly the binary relation on Z^2 depicted in Figure 2.2.5.

(ii) Investigate whether or not the definition you have provided in Exercise 2.6 for the Khalimsky adjacency for Z^3 is the same as κ_3. (It should be!)

4.9. Let N be an integer greater than 1. Prove that none of the following digital spaces is topological.

(i) $\left(Z^N, \alpha_N\right)$.

(ii) $\left(Z^N, \delta_N\right)$.

(iii) $\left(Z^N, \beta_s\right)$, where s is any N-dimensional sign function.

(iv) $\left(Z^N, \varepsilon_e\right)$, where e is any N-dimensional direction vector.

(v) $\left(Z^4, \phi\right)$, where ϕ is as defined on page 42.

4.10. Prove that a linearly path connected subset of R^N is a topologically connected set.

4.11. Show that core of any strictly star-convex set H in R^N is an open set in the standard topology and that it is a subset of the topological interior of H.

4.12. Show that if $G \subset H \subset R^N$, then the topological interior of G is a subset of the topological interior of H.

4.13. We call a subset H of R^N a *star-convex set* if there is an element v in H such that, for every w in H, all elements of the line segment joining v to w are in H.

(i) Show that every strictly star-convex set is a star-convex set.

(ii) Show that every convex set is a star-convex set.

(iii) Give an example of a star-convex set R^2 which has a nonempty topological interior and is neither strictly star-convex nor convex.

4.14. For any positive integer N, show that if $G = Z^N$, then the π of (4.4.6) is in fact ω_N.

4.15. For the ϕ defined on page 42 and for any $c \in Z^4$, show that the topological interior of the union of the Voronoi neighborhoods in Z^4 of all those d which are ϕ-adjacent to c is a topologically connected set (in the standard topology of R^4).

4.16. For any positive integer N and for any Delone grid G in R^N, show that $\{\, N_G(g)\,|\,g \in G \,\}$ is a polyhedral tessellation of R^N.

4.17. Prove that every convex polyhedron in R^N is a convex set in R^N.

4.18. Prove that the following two digital spaces are isomorphic:

(i) (M, ρ), where M is the set of closed squares defined by (4.4.8)-(4.4.11) and ρ is the adjacency defined by (4.4.1).

(ii) (V, π), where V is the set of hexagons indicated in Figure 2.2.3 and π is edge-adjacency.

5
Binary Pictures

"I felt that these pictures had something to say to me that was
important for me to know, but I could not tell what it was."

W. S. Maugham, *The Moon and Sixpence*, Chapter XLII.

5.1. Digital Pictures

After our brief excursion into matters which had to do with topology in the classical
sense, we return to our main topic: the geometry of digital spaces. In fact, this is not
quite correct; we return to digital spaces, but what we do with them in this chapter may be
considered a departure from "geometry."

What we mean by this is the following. In the previous two chapters we have dealt with
things such as surfaces in digital spaces, interiors and exteriors of such surfaces, connectedness
of subsets of digital spaces, and the relationship of such notions to geometrical concepts in
classical topology. One can reasonably claim that all this has been pure "geometry." In this
chapter we move toward what may be considered "analysis" of digital spaces because, in
order to pass from a digital space to a digital picture, we need to introduce a "function" over
the digital space and, in a general sense, *analysis* is the study of functions. An example of
such a function was introduced at the very beginning of this book in the assignment of gray
values to pixels, shown in Figure 1.1.1.

We define a *digital picture* over the digital space (V, π) as a triple (V, π, f), where f is
a function whose domain is V. The choice of topics in this book is biased toward the study
of *binary pictures*, defined as digital pictures in which the range of f is the two-element set
$\{0, 1\}$. We refer to those spels which map into 0 under f as *0-spels* and to those spels which
map into 1 under f as *1-spels*. On occasion, we will restrict our attention further to *finite
pictures*, defined as binary pictures in which the set of 1-spels is finite. (In the terminology
introduced in Chapter 1, we may think of the 1-spels as those which are occupied by "sugar
cubes." In a finite picture the number of sugar cubes is finite.) It is a noteworthy aspect of
our theory that many of our most interesting results apply to binary pictures in general (i.e.,
they do not require that the pictures be finite).

Before plunging into our discussion of binary pictures, we introduce some mathematical
terminology which will be used in this chapter and throughout the rest of the book. As in

Section 1.4, let M be any set, ρ be a binary relation on M, and let A be any subset of M. We define

$$R_\rho(A) = \{\, d \mid (c,d) \in \rho, \text{ for some } c \in A \,\} \tag{5.1.1}$$

and, for any nonnegative integer l,

$$L_\rho^l(A) = \begin{array}{ll} A, & \text{if } l = 0, \\ L_\rho^{l-1}(A) \cup R_\rho\big(L_\rho^{l-1}(A)\big), & \text{if } l > 0. \end{array} \tag{5.1.2}$$

Clearly, ρ is a finitary binary relation if (and only if), for every c in M, $R_\rho(\{c\})$ is finite. (As usual, $\{c\}$ denotes the "singleton" set containing the sole element c.) We will use the fact that if ρ is finitary and A is finite, then $R_\rho(A)$ is finite and $L_\rho^l(A)$ is finite for any nonnegative integer l. We note also that, for arbitrary c and d in M, c is ρ-connected to d if, and only if, d is in $L_\rho^l(\{c\})$ for some nonnegative integer l.

5.2. Fuzzy Segmentation

Digital pictures that we are likely to come across in practice (such as those in Figure 1.1.1) are unlikely to be binary. In this section we give some justification to our choice that, in spite of the fact expressed in the previous sentence, in this book we concentrate on binary pictures. The essential reason is that, for many practical purposes, a digital picture is first turned into a binary picture in the process of either displaying or analyzing the information in the digital picture that we consider essential for the practical task at hand. We have seen examples of this in Chapter 1. In Section 1.2 we discussed distinguishing between "bone" voxels and "not bone" voxels prior to producing computer graphic displays, such as those in Figure 1.1.3. Later in the chapter we talked about identifying "that set of grid points which are in cardiac muscle" for the purpose of estimating the volume of the left ventricle or of the whole heart. In either of these cases, we created (either explicitly or implicitly) a binary picture: bone versus not bone or cardiac muscle versus not cardiac muscle.

The process referred to in Section 1.2 for identifying bone voxels is thresholding. Now we make the nature of this process precise. Thresholding is applicable to *real-valued pictures*, which are digital pictures (V, π, f), with $f : V \to R$. (Real-valued pictures include those, such as the ones shown in Figure 1.1.1, in which the range of values for f is over the integers from 0 to 255. Such pictures are quite typical in practice.) For any $t \in R$, the *t-level set* of the real-valued picture (V, π, f) is defined as the set of spels $\{c \mid f(c) \geq t\}$. Each such t gives rise to a *thresholded picture*, which is the binary picture (V, π, f_t) with

$$f_t(c) = \begin{array}{ll} 1, & \text{if } f(c) \geq t, \\ 0, & \text{if } f(c) < t. \end{array} \tag{5.2.1}$$

We refer to the process of turning a digital picture into a binary picture as *segmentation*. Thresholding is a commonly used method of segmentation. In fact, in much of the medical literature there is even an identification of the two concepts: many writers seem to think that thresholding is the only method for segmentation. This is, of course, not so. Thresholding is not only not the unique method for segmentation, it is very often not the most appropriate method. Examples of this are if there is an overall "shading" in the picture (i.e., the interesting information is superimposed on a background whose darkness slowly varies, as is often the case in many physically obtained digital pictures) and if what distinguishes the object of interest is not the exact values assigned to the individual grid points but rather some textural property. We illustrate below that in both these cases thresholding fails to provide satisfactory segmentation, but it can be achieved by an alternative approach.

The alternative approach that we discuss in detail in this section is the so-called *fuzzy segmentation*. It is definitely not the only possible alternative to thresholding. We have chosen it as our flagship example of turning real-valued pictures into binary pictures mainly because it fits in very smoothly with the theory of digital geometry presented in this book. In particular, we will see that fuzzy segmentation can automatically provide us with Jordan surfaces, which have all the desirable properties discussed in detail in Chapter 3.

A *fuzzy spel affinity* on a digital space (V, π) is a function $\psi : V^2 \to [0, 1]$ (i.e., a function which assigns to each ordered pair of spels a real value not less than 0 and not greater than 1), such that

(i) for all $c \in V$, $\psi(c, c) = 1$, and
(ii) for all $c, d \in V$, $\psi(c, d) = \psi(d, c)$.

The intuitive idea is that the value of the fuzzy spel affinity indicates how close the relationship is (in some sense relevant to the purpose of segmentation) between the spels in question. In other words, it indicates our certainty that upon segmentation the two spels both should be in the same class (e.g., they should either both be identified as bone or both be identified as not bone), with the value 1 indicating absolute certainty. The kind of things that might be considered in defining a fuzzy spel affinity include the nearness of the spels to each other, the similarity of the values assigned to them in a real-valued picture to be segmented, or the similarity of the variance of these assigned values in the immediate neighborhoods of the two spels. For the definition to be useful for segmentation in a specific application, it will usually have to take into consideration the overall purpose of performing the segmentation. We will return to this point below and will give a specific example.

The definition of fuzzy spel affinity is, on purpose, quite general. In particular, there is no requirement of transitivity: it is possible to define a fuzzy spel affinity so that c and d have a high fuzzy spel affinity, d and e have a high fuzzy spell affinity, and yet c and e have a low fuzzy spel affinity. This makes fuzzy spel affinity by itself not immediately useful for segmentation. (It cannot be that both c and d are identified with the same type of tissue and d and e are identified with the same type of tissue, and yet c and e are identified with different types of tissue.) To do segmentation, we first associate with the fuzzy spel affinity ψ a *fuzzy connectedness* function $\mu_\psi : V^2 \to [0, 1]$, defined by

$$\mu_\psi(c, d) = \max_{\substack{\langle c^{(0)}, c^{(k)}, \cdots, c^{(K)} \rangle \\ c = c^{(0)}, \, d = c^{(K)}}} \quad \min_{1 \leq k \leq K} \psi(c^{(k-1)}, c^{(k)}) . \tag{5.2.2}$$

Clearly, $0 \leq \psi(c, d) \leq \mu_\psi(c, d) \leq 1$ (the second inequality follows from the choice $\langle c, d \rangle$ for the sequence from c to d) and so, in particular, $\mu_\psi(c, d) = 1$ if $c = d$.

For a thorough understanding, it is worthwhile to give an intuitive discussion of this last definition. In it we use an arbitrary sequence $\langle c^{(0)}, \cdots, c^{(K)} \rangle$ of spels, which is, of course, a V^2-path. We will informally refer to a V^2-path as a *chain*. If we also think of pairs of consecutive elements (i.e., $(c^{(k-1)}, c^{(k)})$) as *links* in the chain and of $\psi(c^{(k-1)}, c^{(k)})$ as the *strength* of such a link, then we may say that the strength of the chain $\langle c^{(0)}, \cdots, c^{(K)} \rangle$ is the strength of its weakest link, i.e., the minimum of $\psi(c^{(k-1)}, c^{(k)})$ over $1 \leq k \leq K$. Then the fuzzy connectedness of the pair of spels (c, d) is the strength of the strongest chain from c to d. The reason why this concept is useful for segmentation comes from the following quite general result.

Theorem 5.2.1. *For any fuzzy spel affinity ψ on a digital space and for $0 \leq t \leq 1$, the binary relation $K_{\psi,t}$ on V defined by*

$$(c, d) \in K_{\psi,t} \Leftrightarrow \mu_\psi(c, d) \geq t \tag{5.2.3}$$

is an equivalence relation.

Proof. $K_{\psi,t}$ is reflexive since $\mu_\psi(c, c) = 1 \geq t$. From (ii) in the definition of a fuzzy spel affinity, it follows that the strength of a chain $\langle c^{(0)}, c^{(1)}, \cdots, c^{(K)} \rangle$ from c to d is the same as the strength of the chain $\langle c^{(K)}, \cdots, c^{(1)}, c^{(0)} \rangle$ from d to c. Hence $\mu_\psi(c, d) = \mu_\psi(d, c)$, which implies the symmetry of $K_{\psi,t}$. Finally, the strength of the maximal strength chain from c to e must be greater than the minimum of the strengths of any chain from c to d and any chain from d to e. Hence, if there are chains of the latter two types such that the strength of each is at least t, then also $\mu_\psi(c, e) \geq t$. This implies the transitivity of $K_{\psi,t}$. $\square$

This theorem says that any fuzzy spel affinity and any threshold t partitions the set of spels. If the threshold is low, then the equivalence classes of the partition will tend to be large. In particular, $t = 0$ results in V being the only equivalence class. Generally speaking, as the threshold increases, so will the number of equivalence classes, and the individual equivalence classes will tend to be smaller. If the fuzzy spel affinity is strictly less than 1 for any two distinct spels, then $t = 1$ will result in each equivalence class being a singleton set. The equivalence classes of $K_{\psi,t}$ are referred to as ψt-*objects* or, in general, as *fuzzy objects*. These objects have some very desirable properties as expressed in the following theorem and its immediate corollary.

Theorem 5.2.2. *For any fuzzy spel affinity ψ on a digital space (V, π), define the spel-adjacency*

$$\kappa = \pi \cup \{(c, d) \,|\, \psi(c, d) > 0\}. \tag{5.2.4}$$

Let O be a ψt-object.

    (i) *O is κ-connected.*

    (ii) *For any spel-adjacency λ in (V, π), if Q is a λ-connected union of π-components of $\overline{O}$, then $\partial(O, Q)$ is a $\kappa\lambda$-Jordan surface.*

Proof. First we note that κ is indeed a spel-adjacency in (V, π) since it contains π and is symmetric (by (ii) in the definition of fuzzy spel affinity). If O is a $\psi 0$-object, then $O = V$, which is π-connected and hence κ-connected. If O is a ψt-object for some $t > 0$, then for any c and d in O, there is a chain $\langle c^{(0)}, \cdots, c^{(K)} \rangle$ c to d such that, for $1 \leq k \leq K$, $\psi(c^{(k-1)}, c^{(k)}) \geq t > 0$ and, consequently $\langle c^{(0)}, \cdots, c^{(K)} \rangle$ is a κ-path. Furthermore, it is a κ-path in O since (for $1 \leq k \leq K$) $\langle c^{(0)}, \cdots, c^{(k)} \rangle$ is a chain of strength not less than t. This completes the proof of (i). Then (ii) immediately follows from Theorem 3.3.3 since O is a nonempty κ-connected subset of V. $\square$

Corollary 5.2.3. *For any fuzzy spel affinity ψ on a digital space (V, π) such that $\psi(c, d) = 0$ if c and d are not proto-adjacent, let O be a fuzzy object.*

(i) *O is π-connected.*

(ii) *If Q is a π-component of $\overline{O}$, then $\partial(O, Q)$ is a Jordan surface.*

In practice, we are probably not interested in all the fuzzy objects that arise from a particular choice of a fuzzy spel affinity. We are more likely to be concerned with a particular spel and ask which other spels belong to the same fuzzy object as the given one. For example, let us assume that we have found a fuzzy spel affinity appropriate for segmenting bone in CT images (there will be more about how to do that below). When confronted with images such as those in Figure 1.1.1, we may just wish to point at a particular displayed pixel which we are pretty sure is bone-containing (probably because of its lightness) and then ask which other voxels in the three-dimensional array belong to the same piece of bone. This is where the "fuzziness" of the fuzzy segmentation comes into play: the fuzzy object containing the given spel is not uniquely determined by the fuzzy spel affinity (as we decrease the threshold, the same spel is a member of a fuzzy object of increasing size). This reflects the typical state of knowledge in a practical application. From our data alone, it is usually impossible to say with absolute certainty which spels belong to the same object as the selected spel. The larger thresholds give us smaller fuzzy objects but also indicate increasing certainty that we are not attaching a spel which in reality would not be judged to be in the same object.

We formalize the intuitive discussion of the previous paragraph as follows. For any fuzzy spel affinity on a digital space (V, π) and any spel o, we define the *connectedness map for o* as the real-valued picture (V, π, f), where $f(c) = \mu_\psi(o, c)$. The usefulness of this concept is reflected in the following theorem.

Theorem 5.2.4. *Let ψ be a fuzzy spel affinity on a digital space (V, π) and t be a real number, $0 \le t \le 1$*

(i) *The ψt-object that contains a spel o is exactly the t-level set of the connectedness map for o.*

(ii) *If c is in the t-level set of the connectedness map for o, then the t-level set of the connectedness map for c is the same as the t-level set of the connectedness map for o.*

Proof. A spel c is in the ψt-object that contains a spel o if, and only if, $\mu_\psi(o, c) \ge t$, which happens if, and only if, t-level set of the connectedness map for o contains c. This proves (i), and (ii) follows from the fact that $K_{\psi,t}$ is symmetric and transitive. $\square$

We can restate this result as follows. First, we can find all fuzzy objects that contain a given spel by generating the connectedness map for that spel and then thresholding this map at various levels. Second, this procedure is robust: if, instead of finding the connectedness map for o and thresholding at t to get a fuzzy object, we find the connectedness map for any other element of this fuzzy object and threshold that at t, then we end up with exactly the same fuzzy object. In view of this, it is desirable to find an efficient algorithm which, **given** a fuzzy spel affinity ψ in a digital space (V, π) with a finite V and an $o \in V$, **finds** the connectedness map for o. We give an example of such an algorithm under the not unreasonable assumption of the condition of the previous corollary (namely, that $\psi(c, d) = 0$ if c and d are not proto-adjacent). The design strategy of this algorithm is an example of *Dynamic Programming* [38].

Dynamic Program for Fuzzy Objects

(1) Initialization:
    a.   Put o into O.
    b.   Set $f(o) = 1$, and set $f(c) = 0$ if $c \neq o$.

(2) Remove an element d from O. For each spel c that is proto-adjacent to d, do the following.
    a.   Set $v = \min\{f(d), \psi(c, d)\}$.
    b.   If $v > f(c)$, then
        •   Put c into O.
        •   Set $f(c) = v$.

(3) Check if O is empty.
    a.   If it is, STOP.
    b.   If it is not, start again at Instruction (2).

Since this is something of an aside to our main development, we do not give the proof of the fact that, after a finite number of steps, the Dynamic Program for Fuzzy Objects will STOP and at that time, for every $c \in V$, $f(c) = \mu_\psi(o, c)$. (This assumes the finiteness of V.) The efficiency of the algorithm is due to the fact that a just discovered chain from o to any spel c which is not stronger than a previously discovered chain from o to c (this means that $v \leq f(c)$ in Instruction (2)b) is not expanded any further. (This is the basic idea of Dynamic Programming.)

The reader may have noticed that, even though our motivation at the beginning of the section was the segmentation of digital pictures, our formal development has been discussing properties of digital spaces, without any assignment of values to the individual spels. The results we have obtained so far in this section are about digital spaces rather than about digital pictures. However, the important application of these results is the segmentation of digital pictures, as we now discuss.

One of the beauties of fuzzy segmentation is that in many applications an appropriate fuzzy spel-affinity can be automatically created by a computer program, based on some minimal information supplied by a user. It is easiest to demonstrate this by a couple of examples.

Consider Figure 5.2.1. The top left image in this figure is a real-valued picture obtained as follows. First, the set of spels V is a subset of H, the hexagonal grid. Only those elements of H which are inside a large hexagon (with 30 pixels on each side) are in V. (Since our computer display screen has a square grid of display points, in our displays we can only approximate the shape of the hexagonal pixels as a small collection of square pixels.) The real values were assigned to the spels by a three-stage process. First, a uniformly increasing shading was put on the whole picture by assigning the value 99 to the top row of hexagonal pixels and increasing the value assigned to each row by one, until we got to the bottom row, which has the value 157 assigned to it. Second, random noise was introduced in the picture by adding to the value of each pixel a randomly selected number from the range [-33, 33]. Finally, an S-shaped region of pixels was selected, and the values of the pixels in this region were increased by 65. Thus we have a shading across the picture whose range is 59, noise added to the picture whose range is 67, and the signal indicating the S-shaped region has the value 65. Not surprisingly, simply thresholding is not a good way to segment the S-shaped region from its background. This is indicated in the binary picture on the top right of Figure 5.2.1, which has been obtained from the real-valued picture on the top left. (For convenience,

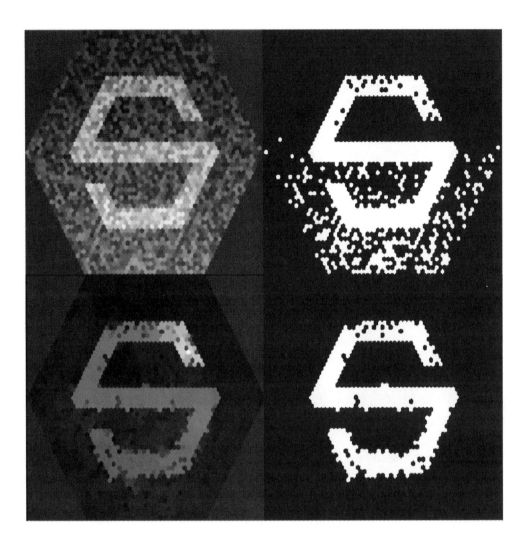

Figure 5.2.1. Example of fuzzy segmentation. Top left: a real-valued picture. Top right: a binary picture obtained by thresholding. Bottom left: a connectedness map. Bottom right: a binary picture obtained by fuzzy segmentation.

we used the thresholding incorporated in our computer. This results in thresholding everything in the square-shaped region of the display pixels, including both the large hexagon and the four surrounding triangles.) Because of the shading, at the selected value of the threshold many spels in the bottom half of the picture have 1s assigned to them, even though they are not in the S-shaped region. This can be ameliorated by raising the threshold, but then we find that the top half of the S-shaped region starts to disappear, since the value of many pixels in it will be below the threshold. There is no single threshold with which we get a reasonable approximation to the S-shaped region in the thresholded binary picture.

An even more fascinating example is illustrated in Figure 5.2.2. Here we first gave to each a hexagonal pixel a randomly selected value from the range [118, 138], and then

Figure 5.2.2. Example of fuzzy segmentation. Top left: a real-valued picture. Top right: a binary picture obtained by thresholding. Bottom left: a connectedness map. Bottom right: a binary picture obtained by fuzzy segmentation.

we added to the pixels in the S-shaped region the additional randomly selected value from the range [-117, 117]. Therefore the real-valued picture on the top left of Figure 5.2.2 is pure noise, and it is only that the noise is more within the S-shaped region than outside it. Thresholding this real-valued picture at its mean value, which is 128, results in the binary image on the top right of Figure 5.2.2. Clearly, this picture gives no information whatsoever as to the location of the S-shaped region.

The automation of finding an appropriate fuzzy affinity function for segmentation of such pictures depends on the following observation: even though a user may not be able to describe the precise nature of the difference between the S-shaped region and its background in the top left images in Figures 5.2.1 and 5.2.2, it is easy to identify some pixels which are definitely in the S-shaped region and some pixels which are definitely in the background. This can be used for gathering some statistical information about the nature of these regions, and then the fuzzy spel affinity can be automatically defined on the basis of such information.

To demonstrate this in our two examples, let us assume (as is the case here and as is also the case in many applications) that the important distinguishing characteristics of regions have to do with the real values assigned to spels in them and also with the likely differences of these values between adjacent spels. More precisely, this translates into the following idea. Given a real-valued picture (V, π, f), we select the fuzzy spel affinity ψ such that, for $c \neq d$,

$$\psi(c, d) = \begin{cases} 0, & \text{if } (c, d) \notin \pi, \\ \frac{1}{2}[g_1(f(c) + f(d)) + g_2(|f(c) - f(d)|)], & \text{otherwise,} \end{cases} \tag{5.2.5}$$

where, for $i \in \{1, 2\}$,

$$g_i(x) = e^{-\frac{(x - m_i)^2}{2\sigma_i^2}}. \tag{5.2.6}$$

We leave it to the reader to check that this ψ is indeed a fuzzy spel affinity on (V, π) and concentrate on the important gap in the definition: what are the values of the m_i and σ_i? In a practical application these values can be obtained as follows. If the user can identify some regions in the real-valued picture which are supposed to belong to the object that should be segmented out from its background (such as regions containing only heart muscle in a CT image), then m_1 and σ_1 can be estimated as the mean and standard deviation, respectively, of $f(c) + f(d)$ over all proto-adjacent spels c and d in the identified regions and m_2 and σ_2 can be estimated as the mean and standard deviation, respectively, of $|f(c) - f(d)|$ over all proto-adjacent spels c and d in the identified regions. This implies that for any pair of proto-adjacent spels c and d whatsoever, their fuzzy spel affinity will be large if both $f(c) + f(d)$ and $|f(c) - f(d)|$ have values typical for the identified regions and will be low if neither $f(c) + f(d)$ nor $|f(c) - f(d)|$ have values typical for the identified regions. In the programs which produced Figures 5.2.1 and 5.2.2, the regions were defined as $L_{\beta_1}^1(o)$, where o is a hexagonally-shaped pixel selected by the user. The user is requested to select two such pixels, providing us with a total of 24 pairs of β_1-adjacent spels in the identified regions (of seven spels each) to calculate the m_i and σ_i.

The bottom left real-valued picture of Figure 5.2.1 is the connectedness map produced by the Dynamic Program for Fuzzy Objects when the two pixels provided by the user were in the S-shaped region. The brightest pixel is the o of the Dynamic Program for Fuzzy Objects, which is also one of the two pixels identified by the user. The bottom right binary picture of Figure 5.2.1 is obtained by thresholding the connectedness map. It is a fuzzy object that is a much better approximation of the S-shaped region than can be obtained by thresholding the original real-valued picture.

The bottom left real-valued picture of Figure 5.2.2 is the connectedness map produced by the Dynamic Program for Fuzzy Objects when the two pixels provided by the user were outside the S-shaped region. The brightest pixel is the o of the Dynamic Program for Fuzzy Objects, which is also one of the two pixels identified by the user. The bottom right binary picture of Figure 5.2.2 is obtained by thresholding the connectedness map. It is a fuzzy object that is a good approximation of the outside of the S-shaped region. In this case, thresholding the original real-valued picture cannot possibly give us any information regarding the region of interest.

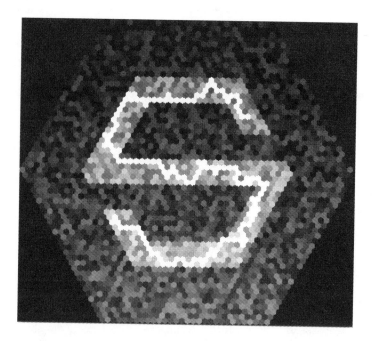

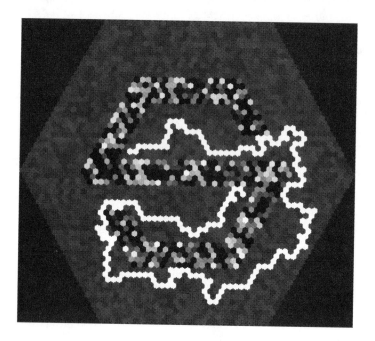

Figure 5.2.3. For both of the real-valued pictures, the user points at two hexagonal pixels. Based on these, a fuzzy spel affinity is calculated as explained in the text. The indicated β_1-path is the strongest possible (for the calculated fuzzy spel affinity) between the two user-selected pixels.

By the way, we mention that a simple alteration of the Dynamic Program for Fuzzy Objects allows us to recover, for any spel c, a π-path of maximal strength from c to o. All we have to do is to store with each spel its predecessor in the chain which has so far been considered strongest, and update this predecessor information when we find a stronger chain to the spel in question. This can have useful applications. For example, it can be used to find a π-path entirely within a region between two user-selected points in that region. This is illustrated in Figure 5.2.3; the indicated β_1-paths are the strongest possible for their respective fuzzy spel affinities, which are defined on the basis of their end points, as explained above.

The principles of fuzzy segmentation have been successfully applied in medical applications; see [44] and its references. It is an efficient, robust, and efficacious technique, which can provide high quality segmented images in spite of the presence of shading and noise. For this reason, in the rest of this book we assume that whatever object we happened to be interested in has already been represented as a connected subset of the 1-spels in a binary image, and we concentrate on the study of binary (rather than real-valued) images in digital spaces. (The reader should not consider this paragraph as a declaration of some ultimate truth. Certainly there will be digital pictures on which the desired segmentation may not be achieved using the fuzzy approach, and there are certainly applications in which alternative approaches to segmentation may well be preferable to fuzzy segmentation.)

5.3. Boundaries in Binary Pictures

Let S be a surface and κ and λ be spel-adjacencies in a digital space (V, π). We say that S is a $\kappa\lambda$-*boundary* in the binary picture (V, π, f), if there is a κ-component O of the set of 1-spels and a λ-component Q of the set of 0-spels such that $S = \partial(O, Q)$. This concept is most important; much of the relevant literature is taken up with the problem of finding $\kappa\lambda$-boundaries in finite pictures for various choices of κ and λ, e.g., [3, 11, 18, 26, 41]. This will also be the topic of Chapter 8.

We emphasize something that will be used without further comment from now on: a $\kappa\lambda$-boundary in a binary picture is always a surface, and so it is a *nonempty* set of surfels. Furthermore, we point out that a $\kappa\lambda$-boundary in a finite picture over a digital space (V, π) with a finitary π is necessarily a finite set of surfels.

Note also that for every surfel (c, d) of a $\kappa\lambda$-boundary in a binary picture it is necessarily the case that c is a 1-spel and d is a 0-spel. This implies immediately that a $\kappa\lambda$-boundary is an antisymmetric surface. Therefore it is possible to develop a theory of $\kappa\lambda$-boundaries in binary pictures based on "unoriented surfels" $\{c, d\}$ since the orientation is automatically provided by the assignment of 0s and 1s. Although this can lead to some mathematically elegant proofs (e.g., [39]), we will not have any occasion to use it in this book.

For reasons much discussed in Chapters 1 and 3, we would like $\kappa\lambda$-boundaries to be $\kappa\lambda$-Jordan. In Theorem 3.3.3 we have seen an example in which a surface defined as a boundary between a κ-connected set of spels and a certain type of λ-connected set of spels is guaranteed to be $\kappa\lambda$-Jordan. Unfortunately, a similar result does not hold in general for $\kappa\lambda$-boundaries.

We have demonstrated this in Figure 1.5.2. In that figure the underlying digital space is (Z^2, ω_2). If we consider those grid points whose Voronoi neighborhoods are shaded in the figure as the 1-spels and all other grid points as the 0-spels, then we see that $O = \{i, k, l, f\}$ is an ω_2-component of the 1-spels and $Q = \{j\}$ is an ω_2-component of the 0-spels. However, $\partial(O, Q) = \{(i, j), (k, j)\}$ is not near-Jordan and hence, by definition, cannot be $\omega_2\omega_2$-Jordan.

As a further demonstration, consider Figure 3.3.1. Define the finite picture (Z^2, ω_2, f) so that all the spels labeled g in Figure 3.3.1 are 1-spels and all other spels are 0-spels. There are eight $\omega_2\omega_2$-boundaries in this binary picture. Four of them contain one surfel each (between one of the spels labeled g and the spel labeled o), and four of them contain three surfels each (between one of the spels labeled g and the three ω_2-adjacent spels labeled e). It is easy to see that none of these eight $\omega_2\omega_2$-boundaries is $\omega_2\omega_2$-Jordan. In fact, for all of them, both the interior and the exterior is the whole of Z^2. On the other hand, all the $\omega_2\alpha_2$-boundaries are $\omega_2\alpha_2$-Jordan (there are four of these, each containing four surfels surrounding one of the spels labeled g), all the $\alpha_2\omega_2$-boundaries are $\alpha_2\omega_2$-Jordan (there are two of these, one containing the four surfels between a spel labeled g and the spel labeled o and the other containing the twelve surfels between the four spels labeled g and the eight spels labeled e), and the one $\alpha_2\alpha_2$-boundary (consisting of all sixteen surfels between a 1-spel and a 0-spel) is $\alpha_2\alpha_2$-Jordan.

Just to make absolutely sure that the reader is on top of all the definitions introduced in Chapter 3 and above, let us dwell on the details of the last statement of the previous paragraph a little longer. In the digital space (Z^2, ω_2), α_2 is a spel-adjacency. The set of spels O which consists of the four spels labeled g in Figure 3.3.1 is clearly α_2-connected and, since it contains all the 1-spels (as defined in the previous paragraph), it is an α_2-component of the set of 1-spels. Its complement Q contains all the spels not labeled g in Figure 3.3.1 and is also clearly α_2-connected. Although it follows already from Theorem 3.3.3 that $\partial(O, Q)$ is $\alpha_2\alpha_2$-Jordan, let us check on the validity of this conclusion based directly on the definitions. A surfel (c, d) is in $\partial(O, Q)$ if, and only if, the spel c is labeled g in Figure 3.3.1 and the spel d is labeled o or e in Figure 3.3.1. Hence $II(\partial(O, Q))$ consists of the four spels labeled g in Figure 3.3.1. Since any ω_2-path from any other spel to one of these spels must cross $\partial(O, Q)$, we see that $\partial(O, Q)$ is near-Jordan and, based on (3.2.2), that $I(\partial(O, Q)) = II(\partial(O, Q)) = O$ is α_2-connected. On the other hand, $IE(\partial(O, Q))$ consists of the spels labeled o or e in Figure 3.3.1, and it is clear that from any spel not labeled g in Figure 3.3.1 there is an α_2-path which does not cross $\partial(O, Q)$ to an element of $IE(\partial(O, Q))$. Hence, $E(\partial(O, Q)) = Q$ is also α_2-connected, which completes the detailed demonstration that $\partial(O, Q)$ is $\alpha_2\alpha_2$-Jordan.

Belaboring this issue a bit further, we wish to emphasize that two very different types of conditions need to be satisfied by a $\kappa\lambda$-boundary S in a digital picture before it can be called a $\kappa\lambda$-Jordan surface. One is that S has to be near-Jordan. To check this, we have to be sure that every π-path from the immediate interior of S to the immediate exterior of S crosses S; this condition is expressed in terms of the proto-adjacency π of the underlying digital space rather than in terms of the spel-adjacencies κ and λ. Now if we look again at the previous three paragraphs, we see that in those cases where we have found that a $\kappa\lambda$-boundary S is not a $\kappa\lambda$-Jordan surface, it has been always because S happened not to be near-Jordan. The other conditions for S to be a $\kappa\lambda$-Jordan surface are that its interior be κ-connected and its exterior be λ-connected. These conditions were satisfied in all the examples above, even in those cases in which the $\kappa\lambda$-boundary S was not a $\kappa\lambda$-Jordan surface. (In fact, in the examples given above, whenever the $\kappa\lambda$-boundary S was not a $\kappa\lambda$-Jordan surface, both the interior and the exterior of S were the whole of Z^2 and so were ρ-connected for any spel-adjacency ρ.) It turns out that this is not accidental. We show below (in Theorem 5.3.7) that, under some quite mild conditions, the interior of a $\kappa\lambda$-boundary is guaranteed to be κ-connected and its exterior is guaranteed to be λ-connected.

A major aim of this book is to provide tools with which we can answer questions of the following type: "Given spel-adjacencies κ and λ in a digital space (V, π), is it the case that every $\kappa\lambda$-boundary in every binary (respectively, in every finite) picture (V, π, f) is $\kappa\lambda$-Jordan?" Much of the rest of this chapter has been developed to be useful when we try to answer such questions.

Lemma 5.3.1. *If κ and λ are spel-adjacencies in a digital space (V, π) and S is a $\kappa\lambda$-Jordan surface in that space, then there exists a binary picture (V, π, f) in which S is the unique $\kappa\lambda$-boundary.*

Proof. By Theorem 3.3.1, since S is a $\kappa\lambda$-Jordan surface in (V, π), there exists a κ-connected subset O of V such that $\overline{O}$ is λ-connected and $S = \partial(O, \overline{O})$. Let f be defined so that elements of O are 1-spels and elements of $\overline{O}$ are 0-spels. Clearly, S is the unique $\kappa\lambda$-boundary in (V, π, f). $\square$

Theorem 5.3.2. *Let κ and λ be spel-adjacencies in a digital space (V, π). A surface S is a $\kappa\lambda$-boundary in some binary picture over (V, π) if, and only if, $\widetilde{S}$ is a $\lambda\kappa$-boundary in some binary picture over (V, π).*

Proof. By Theorem 3.3.7(i), it is sufficient to prove that if S is a $\kappa\lambda$-boundary in some binary picture (V, π, f), then $\widetilde{S}$ is a $\lambda\kappa$-boundary in some binary picture $(V, \pi, \tilde{f})$. It follows from Theorem 3.3.7(ii) that this will be the case if we define $\tilde{f}$ by $\tilde{f}(c) = 1 - f(c)$, for all c in V. $\square$

The same result does not hold in general if we replace "binary" by "finite" in its statement. For example, consider (Z^1, ω_1) with κ chosen as the set of all pairs of spels and λ chosen as ω_1. If we define the finite picture (Z^1, ω_1, f) by $f(i) = 1$ if, and only if, $i = 0$, then one of the two resulting $\kappa\lambda$-boundaries is $S = \{(0, 1)\}$. In this case, $\widetilde{S} = \{(1, 0)\}$. It is easy to see that this cannot be a $\lambda\kappa$-boundary in any finite picture over (Z^1, ω_1). However, we do have the following result.

Theorem 5.3.3. *Let κ and λ be spel-adjacencies in a digital space (V, π), such that κ is finitary. If a surface S is a $\kappa\lambda$-boundary in some finite picture over (V, π), then $\widetilde{S}$ is a $\lambda\kappa$-boundary in some finite picture over (V, π).*

Proof. If S is a $\kappa\lambda$-boundary in some finite picture (V, π, f), then there is a finite κ-component O of the set of 1-spels and a λ-component Q of the set of 0-spels such that $S = \partial(O, Q)$. This S is finite and nonempty, and therefore so are $II(S)$ and $IE(S)$. Let Q' be a finite λ-connected subset of Q that contains $IE(S)$. (Such a Q' exists; we can construct it as the union of the elements of λ-paths in Q which connect pairs of elements of $IE(S)$.) Note that $S = \partial(O, Q')$ and so $\widetilde{S} = \partial(Q', O)$. We define the finite set $B = L^1_\kappa(O \cup Q')$ and the finite picture $(V, \pi, \tilde{f})$ by $\tilde{f}(c) = 1 - f(c)$, if c is in B, and $\tilde{f}(c) = 0$, otherwise.

Now we show that O is a κ-component of the set of 0-spels of $(V, \pi, \tilde{f})$. First of all, since O is a subset of B, every spel in O is a 0-spel of $(V, \pi, \tilde{f})$. Also, O is κ-connected. Therefore all we need to do to complete this part of the proof is to show that any spel d which is κ-connected in the set of 0-spels of $(V, \pi, \tilde{f})$ to a spel c in O is in fact in O. We do this by induction on the κ-path $\langle c^{(0)}, \cdots, c^{(K)} \rangle$ connecting c to d in the set of 0-spels of

$(V, \pi, \tilde{f})$. Since $c^{(0)}$ is c, it is in O. For $1 \leq k \leq K$, assume that $c^{(k-1)}$ is in O. Since $c^{(k)}$ is κ-adjacent to $c^{(k-1)}$, it must be in B, by the definition of B. From this and from the fact that it is a 0-spel of $(V, \pi, \tilde{f})$, it follows that $c^{(k)}$ is a 1-spel of (V, π, f) and so it must be in O.

Now define Q'' as the λ-component in the set of 1-spels of $(V, \pi, \tilde{f})$ that contains Q'. Recall that Q' is a subset of both Q and B (so its elements are 1-spels of $(V, \pi, \tilde{f})$) and that it is λ-connected. Now we prove that Q'' is a subset of Q. Since Q is a λ-component of the set of 0-spels of (V, π, f), it follows that any λ-connected set of 0-spels of (V, π, f) is either contained in Q or is disjoint from Q. This implies that any λ-connected set of 1-spels of $(V, \pi, \tilde{f})$ is either contained in Q or is disjoint from Q, because every 1-spel of $(V, \pi, \tilde{f})$ is a 0-spel of (V, π, f). So, since Q'' is a λ-connected set of 1-spels of $(V, \pi, \tilde{f})$ and is not disjoint from Q, it is contained in Q.

Let $S' = \partial(Q'', O)$. Then S' is a $\lambda\kappa$-boundary in the finite picture $(V, \pi, \tilde{f})$. Furthermore, since $Q' \subset Q'' \subset Q$, we have $\widetilde{S} = \partial(Q', O) \subset \partial(Q'', O) \subset \partial(Q, O) = \widetilde{S}$. It follows that $\widetilde{S} = S'$ is a $\lambda\kappa$-boundary in the finite picture $(V, \pi, \tilde{f})$. $\square$

Note that a consequence of this theorem is that if both κ and λ are finitary spel-adjacencies in a digital space (V, π), then a surface S is a $\kappa\lambda$-boundary in some finite picture over (V, π) if, and only if, $\widetilde{S}$ is a $\lambda\kappa$-boundary in some finite picture over (V, π). The restriction to finitary spel-adjacencies is not a serious one in practice; all spel-adjacencies that have been studied with an application in mind are finitary.

Having proved these results which shed some light on the nature of $\kappa\lambda$-boundaries, we now work towards a much more significant result. Unfortunately, however, this result does not hold for all spel-adjacencies in an arbitrary digital space. So before stating it, we need to specify the type of spel-adjacencies for which it holds.

First we need a rather technical, but very important, definition (based on an idea of T. Y. Kong). Let ρ be a binary relation on a set M. We say that $\langle c^{(0)}, \cdots, c^{(K)} \rangle$ is a ρ-*tight sequence* of elements in M if either $K = 0$ or, for $0 \leq k \leq K$, either $(c^{(0)}, c^{(k)}) \in \rho$ or $(c^{(k)}, c^{(K)}) \in \rho$ (or possibly both). Now let ρ be a spel-adjacency in a digital space (V, π). We call ρ a *tight spel-adjacency* if

$$(c, d) \in \rho \Rightarrow (\text{there exists a } \rho\text{-tight } \pi\text{-path from } c \text{ to } d). \tag{5.3.1}$$

Trivially, π and V^2 are tight spel-adjacencies in (V, π). The binary relations α_2, ω_2, β, γ, and δ on Z^2 (as defined in Chapters 1 and 2) are tight spel-adjacencies in (Z^2, ω_2). On the other hand, they are not tight in (Z^2, χ), as defined on page 59. This is because two pixels are α_2-, ω_2-, β-, γ-, and δ-adjacent as long as one of them is immediately above the other one, but if they are more than two pixels away from the vertical axis, then there is no α_2-, ω_2-, β-, γ-, or δ-tight χ-path from one to the other.

As a more complicated example, consider the not finitary binary relation ν on Z^2 defined on page 67. This is a tight spel-adjacency in both (Z^2, χ) and in (Z^2, ω_2), as we show now. If a pair of pixels is in ν, then either they are in the same row or one is immediately above the other (see Figure 3.3.3). In the former case, the horizontal χ-path from one to the other is ν-tight. In the latter case, the χ-path which goes horizontally from the first pixel to the vertical axis, then makes a single vertical step, and then goes horizontally to the second pixel is ν-tight. Since every χ-path is also an ω_2-path, our argument is complete.

Now we discuss the reasonableness of believing that condition (5.3.1) would be satisfied by spel-adjacencies of practical interest. At a general heuristic level, we can justify (5.3.1) by considering the unreasonableness of the alternative. If (5.3.1) does not hold, then there exist spels c and d which are ρ-adjacent, but for all π-paths from c to d there is some intermediate spel which is not ρ-adjacent to either c or d. This means that ρ "jumps over" spels between c and d, an intuitively undesirable property of something that is to be called an "adjacency."

Much more specifically, consider local spel-adjacencies in the digital spaces (Z^N, ω_N), as defined in Chapter 4. Our geometrical interpretation of a local adjacency was that the Voronoi neighborhoods (in the N-dimensional euclidean space) of two adjacent spels have a nonempty intersection. It turns out that we need only assume locality to ensure that an arbitrary spel-adjacency in either of the spaces (Z^2, ω_2) or (Z^3, ω_3) is tight (i.e., satisfies (5.3.1)). This is a significant statement, since some surveys on digital topology restrict their attention nearly entirely to the digital spaces (Z^2, ω_2) and (Z^3, ω_3) and deal with local spel-adjacencies in those spaces almost exclusively; e.g., [24, 25]. (In fact, for an arbitrary positive integer N, α_N, δ_N, and κ_N are local trivially from the definitions, as are β_s and ε_e; see Exercise 4.7.) Here we prove a slightly more general result.

Let ρ be a spel-adjacency in a digital space (V, π), and let l be a nonnegative integer. We say that ρ is *l-limited* if, for every c in V,

$$R_\rho(\{c\}) \subset L_\pi^l(\{c\}) . \tag{5.3.2}$$

The proto-adjacency itself is 1-limited. An example of a 3-limited, but not 2-limited, spel-adjacency in (Z^2, ω_2) is δ (see page 51). More importantly, it is easy to see that if ρ is a local spel-adjacency in (Z^2, ω_2), then ρ is 2-limited (and hence 3-limited) and if ρ is a local spel-adjacency in (Z^3, ω_3), then ρ is 3-limited. A spel-adjacency which is not l-limited for any l is said to be *unlimited*.

Theorem 5.3.4. *Every 3-limited spel-adjacency in a digital space is tight.*

Proof. Let ρ be a 3-limited spel-adjacency in a digital space (V, π), and consider ρ-adjacent spels c and d. Since (5.3.2) is satisfied with $l = 3$, we see from the definitions of $R_\rho(A)$ and of $L_\rho^l(A)$ that there exists a π-path $\langle c^{(0)}, \cdots, c^{(K)} \rangle$ with $K \leq 3$ from c to d. Now we show that this π-path is ρ-tight. This is trivially so if $K = 0$. Otherwise, observe that $(c^{(0)}, c^{(1)}) \in \rho$ and $(c^{(K-1)}, c^{(K)}) \in \rho$, since $\pi \subset \rho$ by the definition of a spel-adjacency. This, combined with the facts that $(c^{(0)}, c^{(K)}) \in \rho$ and that $1 \leq K \leq 3$, completes the proof of the ρ-tightness of $\langle c^{(0)}, \cdots, c^{(K)} \rangle$. $\square$

This theorem, combined with the paragraph before it, shows that every local spel-adjacency in (Z^2, ω_2) or in (Z^3, ω_3) is tight. It is easy to illustrate that the corresponding result is not valid for (Z^N, ω_N) with $N \geq 4$. Nevertheless, as we will show now, the spel-adjacencies that were of sufficient interest to have been explicitly defined for such spaces are all tight!

In fact, we will show a stronger result, one which will not be needed in this section but will be useful later on. We say that a spel-adjacency ρ in the digital space (V, π) is a *very tight spel-adjacency* if

$$(c, d) \in \rho \Rightarrow (\text{ there exists a } \pi\text{-path } \langle c^{(0)}, \cdots, c^{(K)} \rangle \text{ from } c \text{ to } d,$$
$$\text{such that } (c, c^{(k)}) \in \rho, \text{ for } 1 \leq k \leq K) . \tag{5.3.3}$$

We leave it to the reader to show that

(i) ρ is very tight if, and only if, $L_\rho^1(\{c\})$ is π-connected for all spels c;

(ii) if ρ is very tight, then ρ is tight;

(iii) if ρ is 2-limited, then ρ is very tight.

The converse of (ii) does not hold; for example, the δ defined on page 51 is a tight but not a very tight spel-adjacency in (Z^2, ω_2). Being 2–limited is a sufficient but not a necessary condition for being very tight: the unlimited spel-adjacency ν illustrated in Figure 3.3.3 is clearly very tight in (Z^2, ω_2). More importantly, now we show that the condition of being very tight does not exclude adjacencies of practical interest.

Theorem 5.3.5. *For every positive integer N, the following spel-adjacencies are very tight (and hence tight) in (Z^N, ω_N): α_N, δ_N, κ_N, ω_N, ε_e (for any N-dimensional direction vector e) and, when $N \geq 2$, β_s (for any N-dimensional sign function s).*

Proof. Since δ_N, ω_N and β_s are 2-limited, they are very tight; see (iii) above. For the others we have to create explicitly a π-path which has the property stated in (5.3.3). We do this only for ε_e. The treatment of α_N and κ_N is analogous.

If $(c, d) \in \varepsilon_e$, then there are some, say K, values of n for which $c_n \neq d_n$ and, consequently, $|c_n - d_n| = |e_n| = 1$. We define the required path $\langle c^{(0)}, \cdots, c^{(K)} \rangle$ in (Z^N, ω_N) from c to d by changing one-by-one the nth component for those ns for which $c_n \neq d_n$ from c_n to d_n. $\square$

This theorem shows (especially if we consider Exercise 5.9) that the set of tight (or even of very tight) spel-adjacencies contains all the adjacencies which have been used to motivate our theory, and so there is essentially no practical loss in restricting our attention to them. This makes the next theorem (and other results which follow) much more significant than it would be otherwise.

Theorem 5.3.6. *Let (V, π, f) be a binary picture and κ and λ be tight spel-adjacencies in (V, π). If O is a κ-component of the set of 1-spels, Q is a λ-component of the set of 0-spels, and S is the $\kappa\lambda$-boundary $\partial(O, Q)$, then $O \subset I(S)$ and $Q \subset E(S)$.*

Proof. We prove only that $O \subset I(S)$, since the proof that $Q \subset E(S)$ is strictly analogous. Let $d \in O$, $c \in II(S)$ $(\subset O)$, and $\langle c^{(0)}, \cdots, c^{(K)} \rangle$ be a κ-path in O from c to d. We prove by induction that, for $0 \leq k \leq K$, $c^{(k)} \in I(S)$. This suffices to complete our proof.

Clearly, $c^{(0)} \in I(S)$. Now assume that the same is true for $c^{(k-1)}$, for some k, $1 \leq k \leq K$. Since $(c^{(k-1)}, c^{(k)}) \in \kappa$, there exists a κ-tight π-path $\langle e^{(0)}, \cdots, e^{(T)} \rangle$ from $c^{(k-1)}$ to $c^{(k)}$. Now we prove by induction that, for $0 \leq t \leq T$,

$$e^{(t)} \in I(S) \cup Q. \tag{5.3.4}$$

Since $e^{(T)} = c^{(k)}$ is in O and so cannot be in Q, this shows that $c^{(k)} \in I(S)$ and so completes our proof.

If $t = 0$, then (5.3.4) is satisfied since $e^{(0)} = c^{(k-1)}$ is in $I(S)$ by the hypothesis for the induction on k. Now assume that (5.3.4) is satisfied for some $t < T$. In case $e^{(t)} \in I(S)$, either $e^{(t+1)} \in I(S)$ (in which case we are done) or $(e^{(t)}, e^{(t+1)}) \in S$ (see (3.2.2)), and so $e^{(t+1)} \in Q$ (and again we are done). In case $e^{(t)} \in Q$, either $e^{(t+1)}$ is a 0-spel and therefore is in Q since $\pi \subset \lambda$ (in which case we are done) or $e^{(t+1)} \in O$, since the π-path $\langle e^{(0)}, \cdots, e^{(T)} \rangle$ from $c^{(k-1)}$ to $c^{(k)}$ is κ-tight and both $c^{(k-1)}$ and $c^{(k)}$ are in O. In this latter case, $(e^{(t+1)}, e^{(t)}) \in S$, and so $e^{(t+1)} \in I(S)$ (and again we are done). $\square$

The proof of this theorem made essential use of the definition of tightness of spel-adjacencies. This brings us to a discussion of the second reason why tightness is such an essential notion in our theory. We have already discussed (just before the statement of the previous theorem) that there is no practical loss in restricting our attention to tight spel-adjacencies. Now we indicate why it is important that we do so: it is because intuitively desirable properties (such as the ones stated in the conclusion of the previous theorem) would no longer be guaranteed if we did not restrict our study to tight spel-adjacencies. To demonstrate this, we give an example which illustrates that Theorem 5.3.6 fails to hold without the assumption that both κ and λ are tight.

Our example is over the set of integers Z. For each positive integer i, we define the adjacency ν_i on Z by

$$(c, d) \in \nu_i \iff |c - d| = 1 \text{ or } |c - d| = i . \tag{5.3.5}$$

Clearly, ν_i is a spel-adjacency in (Z, ν_1), which is obviously isomorphic to (Z^1, ω_1), and (by Theorem 5.3.4) ν_1, ν_2 and ν_3 are in fact tight spel-adjacencies. On the other hand, it is clearly the case that ν_4 is not a tight spel-adjacency in (Z, ν_1).

Now consider the binary picture (Z, ν_1, f) in which the only 1-spels are 0, 2 and 4. Then $O = \{0, 4\}$ is a ν_4-component of the set of 1-spels and $Q = \{1\}$ is a ν_1-component of the set of 0-spels (see Figure 5.3.1). In this case, $S = \partial(O, Q) = \{(0, 1)\}$ and $I(S) = \{j \mid j \leq 0\}$. It follows that $O \not\subset I(S)$.

Theorem 5.3.7. *Let κ and λ be tight spel-adjacencies in a digital space (V, π). If S is a $\kappa\lambda$-boundary in a binary picture (V, π, f), then $I(S)$ is κ-connected and $E(S)$ is λ-connected.*

Proof. Let $S = \partial(O, Q)$, for some κ-component O of the set of 1-spels and some λ-component Q of the set of 0-spels. Then, by Theorem 5.3.6, $O \subset I(S)$ and $Q \subset E(S)$ and so, by Lemma 3.3.2, $I(S)$ is κ-connected and $E(S)$ is λ-connected. $\square$

The importance of this theorem cannot be overemphasized. Our desire is that $\kappa\lambda$-boundaries should be $\kappa\lambda$-Jordan surfaces. The theorem we have just proved says that, as long as κ and λ are tight spel-adjacencies, the κ-connectedness of the interior and the λ-connectedness of the exterior of a $\kappa\lambda$-boundary in a binary picture are automatically satisfied. The only thing that we have to worry about is ensuring that the $\kappa\lambda$-boundary is near-Jordan.

The assumption that both κ and λ are tight is also essential for the proof of this theorem (see Exercises 5.11 and 5.12). This further demonstrates why the set of tight spel-adjacencies is the appropriate domain of interest for our study.

Figure 5.3.1. The heavily shaded spels represent O, a ν_4-component of the set of 1-spels. The lighter shaded spel represents Q, a ν_1-component of the set of 0-spels. The $\nu_4\nu_1$-boundary $\partial(O, Q)$ is represented by the heavy line; its interior consists of all the spels to the left of the heavy line and its exterior consists of all the spels to the right of the heavy line. The interior does not contain all of O.

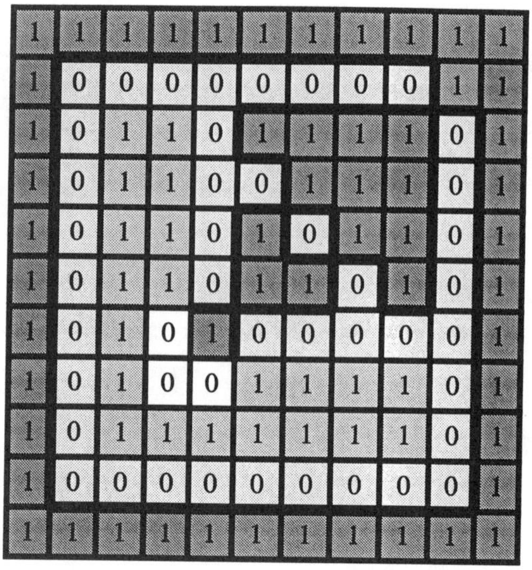

Figure 5.3.2. Assignment of 0s and 1s in the finite picture illustrating Theorem 5.3.7.

Note also that in Theorem 5.3.7 S is not required to be near-Jordan. To demonstrate its power, now we give an example of a $\beta\gamma$-boundary (recall Figure 2.2.1) S in (Z^2, ω_2), for which $I(S) \cap E(S) \neq \emptyset$ (and so, by Lemma 3.2.2, S is not near-Jordan), but nevertheless $I(S) \neq Z^2$ and $E(S) \neq Z^2$. In spite of these properties, the interior of S is β-connected and the exterior of S is γ-connected, as indeed they have to be according to Theorem 5.3.7.

Consider the assignment of 0s and 1s to elements of Z^2 which is shown in Figure 5.3.2. All spels not shown in the figure are 0-spels. There are two β-components of the set of 1-spels, these are assigned different shades. The heavier shade is assigned to the β-component used in the definition of the $\beta\gamma$-boundary. There are three γ-components of the set of 0-spels. The one used in the definition of the $\beta\gamma$-boundary is assigned a very light shade, the second consists of the three spels in the figure which are not shaded at all, and the third consists of all the spels not shown in the figure. The surfels in the $\beta\gamma$-boundary are indicated by the bold edges; the orientation is always from the 1-spel to the 0-spel.

By the definition of the interior of a surface, we see that every spel in Z^2 except one is in the interior of S. The single exception is the 0-spel in the fifth row and seventh column. On the other hand, the exterior of S consists of all the spels in the two finite γ-components of the set of 0-spels, together with the spels in the β-component of the set of 1-spels which was not used for the definition of the $\beta\gamma$-boundary and a single 1-spel in the component used for the definition of the $\beta\gamma$-boundary (the one in the seventh row and fifth column). Thus we see that $I(S) \neq Z^2$ and $E(S) \neq Z^2$, but the intersection of the interior and the exterior is nonempty. In fact, it contains both 0-spels and 1-spels. It is easy to check that the interior is β-connected and the exterior is γ-connected, as they should be according to Theorem 5.3.7.

Theorem 5.3.8.   *Let κ and λ be tight spel-adjacencies and S be a surface in a digital space (V, π). Then the following two statements are equivalent.*

    (i) *S is a near-Jordan $\kappa\lambda$-boundary in some binary picture (V, π, f).*

(ii) S *is* $\kappa\lambda$-*Jordan.*

Proof. Noting the definition of a $\kappa\lambda$-Jordan surface, we see that "(i) implies (ii)" follows immediately from Theorem 5.3.7 and that "(ii) implies (i)" follows immediately from Lemma 5.3.1. □

5.4. Jordan Pairs of Spel-Adjacencies

In certain digital spaces there exist pairs of tight spel-adjacencies κ and λ such that every $\kappa\lambda$-boundary in every binary picture is near-Jordan. This section discusses the problem of identifying such pairs of tight spel-adjacencies and related problems. The usefulness of this is indicated by Theorem 5.3.8; for such κ and λ, every $\kappa\lambda$-boundary in every binary picture is $\kappa\lambda$-Jordan.

An unordered pair $\{\kappa, \lambda\}$ of tight spel-adjacencies in the digital space (V, π) is said to be a

(i) *strong Jordan pair* for (V, π) if every $\kappa\lambda$-boundary in every binary picture over (V, π) is near-Jordan;

(ii) *Jordan pair* for (V, π) if every finite $\kappa\lambda$-boundary in every binary picture over (V, π) is near-Jordan;

(iii) *weak Jordan pair* for (V, π) if every finite $\kappa\lambda$-boundary and every finite $\lambda\kappa$-boundary in every finite picture over (V, π) is near-Jordan.

Note that these concepts are defined only for tight spel-adjacencies and so if we say that $\{\kappa, \lambda\}$ is a (strong, weak) Jordan pair, then automatically we imply that both κ and λ are tight spel-adjacencies.

First we explain why it is possible to define the first two of these concepts for *unordered* pairs $\{\kappa, \lambda\}$ in an asymmetrical way. Suppose that $\tilde{S}$ is a $\lambda\kappa$-boundary in a binary picture over (V, π). Then, by Theorem 5.3.2, S is a $\kappa\lambda$-boundary in some binary picture over (V, π). So if $\{\kappa, \lambda\}$ is strong Jordan by the definition given above, then S is near-Jordan, and hence, by Theorem 3.3.7(v), $\tilde{S}$ is near-Jordan. So the order of κ and λ does not matter for the definition of a strong Jordan pair. Exactly the same argument can be repeated to justify the definition of a Jordan pair, simply by noting that S is finite if, and only if, $\tilde{S}$ is finite.

Theorem 5.4.3 below shows that no such argument can be given for weak Jordan pairs; that is why their definition is made explicitly symmetrical. However, before getting into the technical details of this, we state and prove a most important theorem. In the next chapter we will show that Claims 1.5.2, 1.6.2, and 2.1.1 are all straightforward consequences of the following result.

Theorem 5.4.1. *Let* (V, π, f) *be a binary picture,* κ *and* λ *be tight spel-adjacencies in* (V, π), O *be a* κ-*component of the set of 1-spels, and* Q *be a* λ-*component of the set of 0-spels such that* $S = \partial(O, Q)$ *is not empty. Further suppose that at least one of the following three conditions is satisfied:*

- $\{\kappa, \lambda\}$ *is a strong Jordan pair;*
- $\{\kappa, \lambda\}$ *is a Jordan pair, and* S *is finite;*
- $\{\kappa, \lambda\}$ *is a weak Jordan pair,* S *is finite, and* (V, π, f) *is a finite picture.*

Then there exist two subsets I and E of V, with the following properties.

(i) $O \subset I$ and $Q \subset E$.

(ii) $\partial(O, Q) = \partial(I, E)$.

(iii) $I \cup E = V$ and $I \cap E = \emptyset$.

(iv) I is κ-connected and E is λ-connected.

(v) *Every π-path from an element of I to an element of E crosses $S = \partial(O, Q)$.*

Furthermore, for any such I and E, it must be the case that $I = I(S)$ and $E = E(S)$.

Proof. By the definitions of a strong Jordan pair, Jordan pair and weak Jordan pair, if any one of the three conditions is satisfied, then S is near-Jordan. t follows Therefore it follows from Theorem 3.2.5 that there exists a nonempty proper subset I of V such that $S = \partial(I, \overline{I})$ and that for this I we have that $I = I(S)$. We see immediately that (iii) above can be satisfied if, and only if, $E = \overline{I}$ and that, in this case, (ii) is also satisfied. Furthermore, from Lemmas 3.2.1 and 3.2.2 (and from the already proven fact that $I = I(S)$), it follows that $E = E(S)$. In view of this, (i) follows from Theorem 5.3.6, (iv) follows from Theorem 5.3.7, and (v) follows from Lemma 3.2.2. $\square$

Having established the importance of the notions of strong Jordan pairs, Jordan pairs, and weak Jordan pairs, now we return to the discussion of the relationship between the last two of them. First we need a technical result.

Lemma 5.4.2. *If κ is an arbitrary spel-adjacency in the digital space (Z^2, ω_2) and ν is the spel-adjacency defined on page 67, then every $\kappa\nu$-boundary in every finite picture over (Z^2, ω_2) is near-Jordan.*

Proof. Let S be a $\kappa\nu$-boundary in a finite picture (Z^2, ω_2, f). By definition, $S = \partial(O, Q)$ for a nonempty and necessarily finite κ-component O of the set of 1-spels and a ν-component Q of the set of 0-spels. In the next paragraph we show that in fact every 0-spel is in Q. This implies that $S = \partial(O, \overline{O})$ (since $S \subset \partial(O, \overline{O})$ and if $(c, d) \in \partial(O, \overline{O})$, then d must be a 0-spel) and so, by Theorem 3.2.5, S is near-Jordan.

Since O is finite, there is a positive integer C such that $C > \|c\|$ for every $c \in O$. Let $E = \{e \in Z^2 \mid \|e\| \geq C\}$; clearly, E is a set of 0-spels. Since (Z^2, ω_2) is a boundable digital space (Theorem 3.2.3), E is ω_2-connected. It is also clear from the definition of ν that, for any 0-spel d, there is an $e \in E$ such that $(d, e) \in \nu$ (see Figure 3.3.3). It follows that the set of 0-spels is ν-connected and, since Q is a ν-component of the set of 0-spels, it must contain all the 0-spels. $\square$

Theorem 5.4.3. *There exists a digital space (V, π) and tight spel-adjacencies κ and λ in (V, π) such that*

(i) *every finite $\kappa\lambda$-boundary in every finite picture over (V, π) is near-Jordan and*

(ii) *there exists a finite $\lambda\kappa$-boundary in a finite picture over (V, π) which is not near-Jordan.*

Proof. Consider the digital space (Z^2, ω_2) and the spel-adjacencies ω_2 and ν. (That ν is a tight spel-adjacency in (Z^2, ω_2) is discussed on page 114.) It follows immediately from Lemma 5.4.2 that every finite $\omega_2\nu$-boundary in every finite picture over (V, π) is near-Jordan.

To complete the proof, consider Figure 3.3.1. Let all the spels labeled g be 1-spels and all other spels be 0-spels, giving rise to a finite picture over (Z^2, ω_2). Clearly, the uppermost of the spels labeled g forms on its own a ν-component of the set of 1-spels (see Figure

3.3.3) and the spel labeled o forms on its own an ω_2-component of the set of 0-spels. The finite $\nu\omega_2$-boundary (consisting of one surfel) between these two components is clearly not near-Jordan. $\square$

In this proof we have used a not finitary spel-adjacency as our choice for λ. In fact, we could not possibly have proved Theorem 5.4.3 using a finitary λ, as can be seen from our next result.

Theorem 5.4.4. *Let (V, π) be a digital space with κ and λ spel-adjacencies in (V, π) such that λ is finitary. If every finite $\kappa\lambda$-boundary in every finite picture over (V, π) is near-Jordan, then every finite $\lambda\kappa$-boundary in every finite picture over (V, π) is near-Jordan.*

Proof. Let $\widetilde{S}$ be a finite $\lambda\kappa$-boundary in a finite picture over (V, π). If λ is finitary, it follows from Theorem 5.3.3 that $\widetilde{\widetilde{S}} = S$ (see Theorem 3.3.7(i)) is a $\kappa\lambda$-boundary in some finite picture over (V, π). It is also clear that S is finite and so, if every finite $\kappa\lambda$-boundary in every finite picture over (V, π) is near-Jordan, then S (and hence $\widetilde{S}$, see Theorem 3.3.7(v)) is near-Jordan. $\square$

It immediately follows from the proof of Theorem 5.4.3 (in which we used a finitary κ) that the converse of the conclusion of the previous theorem (i.e., the second sentence with κ and λ interchanged) does not hold.

We note that a strong Jordan pair is necessarily a Jordan pair and a Jordan pair is necessarily a weak Jordan pair. We postpone till Chapter 7 the task of showing that there are Jordan pairs which are not strong Jordan pairs. Here we show the corresponding result for weak Jordan pairs and Jordan pairs.

Theorem 5.4.5. *There exists a digital space (V, π) and tight spel-adjacencies κ and λ in (V, π) such that $\{\kappa, \lambda\}$ is a weak Jordan pair, but is not a Jordan pair.*

Proof. Consider the digital space (Z^2, ω_2) and the pair $\{\nu, \nu\}$ of tight (see page 114) spel-adjacencies. That this is a weak Jordan pair follows immediately from Lemma 5.4.2.

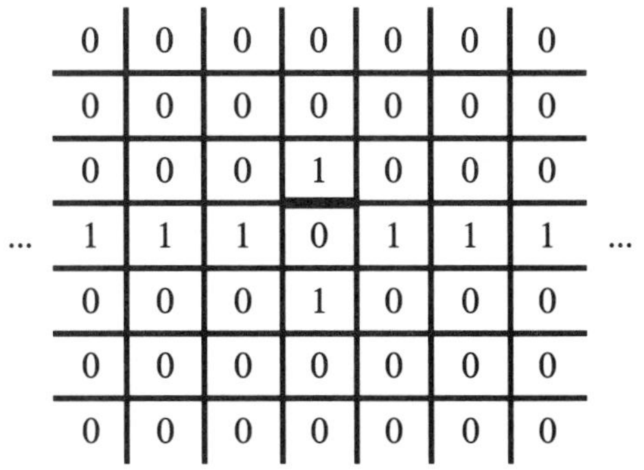

0	0	0	0	0	0	0	
0	0	0	0	0	0	0	
0	0	0	1	0	0	0	
1	1	1	0	1	1	1	
0	0	0	1	0	0	0	
0	0	0	0	0	0	0	
0	0	0	0	0	0	0	

Figure 5.4.1. The binary picture used in the proof of Theorem 5.4.5. All the not-indicated spels of (Z^2, ω_2) are 0-spels with the exception of the (infinitely many) spels in the central row.

On the other hand, consider the binary picture represented in Figure 5.4.1. The surfel indicated by the heavy edge is a $\nu\nu$-boundary, since the spel above it forms a ν-component of the set of 1-spels and the spel below it forms a ν-component of the set of 0-spels (see Figure 3.3.3). This finite $\nu\nu$-boundary is not near-Jordan. $\square$

In spite of this result, the distinction between weak Jordan pairs and Jordan pairs of spel-adjacencies is not very important. To prove Theorem 5.4.5, we had to drag in the rather odd ball spel-adjacency ν. We show below that under some very mild assumptions which are certainly valid for all spel-adjacencies of practical interest, a weak Jordan pair of spel-adjacencies is guaranteed to be a Jordan pair. Thus, for all practical purposes, the concepts of a Jordan pair and of a weak Jordan pair are equivalent.

As we have seen in the discussion following the definition of a $\kappa\lambda$-boundary at the beginning of the previous section, the pair $\{\omega_2, \omega_2\}$ of finitary tight spel-adjacencies in the digital space (Z^2, ω_2) is not a Jordan pair for (Z^2, ω_2), since we can define (based on Figure 3.3.1) a binary picture over (Z^2, ω_2) in which there are finite $\omega_2\omega_2$-boundaries which are not near-Jordan. On the other hand, we claim that $\{\chi, \chi\}$ is a strong Jordan pair for (Z^2, χ), where χ is defined on page 59. To see this, consider any nonempty component O of the set of 1-spels in a binary picture over (Z^2, χ). We distinguish between two possibilities. In the first, O does not contain any spels c for which $c_1 = 0$ (i.e., on the vertical axis). In this case, O has to be a (possibly semi-infinite) horizontal run of 1-spels, of which only the end one(s) may be χ-adjacent to a 0-spel. It is also easy to see that if there happened to be two such 0-spels (because the run of 1-spels is finite), then these two 0-spels could not be χ-connected in the set of 0-spels. Hence, in this case, any resulting $\chi\chi$-boundary contains exactly one surfel and, therefore, as pointed out before, is near-Jordan (see page 60). The other possibility is that O contains at least one spel c for which $c_1 = 0$. In this case, O has to be a (possibly semi-infinite or infinite) vertically contiguous stack of (possibly semi-infinite or infinite) horizontal runs of 1-spels, each one of which contains a spel c for which $c_1 = 0$. Again, one can show that two 0-spels which are both χ-adjacent to some spel in O cannot possibly be χ-connected in the set of 0-spels. Hence, in this case also, any resulting $\chi\chi$-boundary contains exactly one surfel and therefore is near-Jordan.

The strong Jordan pair of this last example is "minimal" in the sense that any spel-adjacency in (Z^2, χ) must contain χ. *The following theorem provides motivation for finding such minimal strong Jordan pairs (or minimal Jordan pairs or minimal weak Jordan pairs).* In its statement (and in the rest of the book) we use a linguistic device, much loved by mathematicians, to express several ideas simultaneously in one sentence. This device looks like this: "this (respectively that or something else)." What is implied by this is that the sentence containing the phrase "this" can be repeated in its entirety by the same sentence, but with "this" replaced by "that" or by "something else." Using this device, the sentence in italics above can be rewritten as "The following theorem provides motivation for finding such minimal strong Jordan pairs (respectively, Jordan pairs or weak Jordan pairs)."

Theorem 5.4.6. *If $\{\kappa, \lambda\}$ is a strong Jordan pair (respectively, a Jordan pair or a weak Jordan pair) for a digital space (V, π) and κ' and λ' are tight spel-adjacencies in (V, π) such that $\kappa \subset \kappa'$ and $\lambda \subset \lambda'$, then $\{\kappa', \lambda'\}$ is also a strong Jordan pair (respectively, a Jordan pair or a weak Jordan pair) for the digital space (V, π).*

Proof. In the proof we assume that $\lambda = \lambda'$. We can do this since the result stated in the theorem will obviously follow from two applications of what we actually show.

Suppose that $\{\kappa, \lambda\}$ is a strong Jordan pair (respectively, a Jordan pair) for a digital space (V, π), κ' is a tight spel-adjacency in (V, π) such that $\kappa \subset \kappa'$, and S is a $\kappa'\lambda$-boundary (respectively, a finite $\kappa'\lambda$-boundary) in some binary picture (V, π, f). Let c be in $II(S)$ and d be in $IE(S)$. We need to show that every π-path from c to d crosses S.

Let O be the κ-component and O' be the κ'-component of the set of 1-spels in (V, π, f) which contain c and let Q be the λ-component of the set of 0-spels in (V, π, f) which contains d. Then $S = \partial(O', Q)$. Since $\kappa \subset \kappa'$, we have $O \subset O'$, and $\partial(O, Q) \subset S$. Furthermore, $\partial(O, Q)$ is not empty, since the fact that c (which is in O) is in $II(S)$ implies that there must be an e in $IE(S)$ (and hence in Q) such that $(c, e) \in \partial(O, Q)$. If S is finite, so is $\partial(O, Q)$. The fact that $\{\kappa, \lambda\}$ is a strong Jordan pair (respectively, a Jordan pair) for a digital space (V, π) implies that the $\kappa\lambda$-boundary $\partial(O, Q)$ is near-Jordan. By Theorem 5.3.6, we also know that c, which is in O, is in the interior of $\partial(O, Q)$ and d, which is in Q, is in the exterior of $\partial(O, Q)$. By Lemma 3.2.2, every π-path from c to d crosses $\partial(O, Q)$ and hence crosses S since S is a superset of $\partial(O, Q)$.

This completes the proof for strong Jordan pairs and Jordan pairs. The proof for weak Jordan pairs is essentially the same, but the argument has to be given both for finite $\kappa'\lambda$-boundaries and for finite $\lambda\kappa'$-boundaries in finite pictures. $\square$

Clearly, this theorem provides us with a powerful tool for obtaining strong Jordan pairs (respectively, Jordan pairs or weak Jordan pairs). For example, from the fact that $\{\chi, \chi\}$ is a strong Jordan pair for (Z^2, χ) (discussed just before the statement of the theorem), it follows that $\{\nu, \nu\}$ is a strong Jordan pair for (Z^2, χ), which is not itself obvious and may at first sight be considered surprising in view of the fact that $\{\nu, \nu\}$ is not even a Jordan pair for (Z^2, ω_2), as can be seen from the proof of Theorem 5.4.5.

It follows from the previous paragraph that for (Z^2, χ) the concepts of a strong Jordan pair and of a Jordan pair are equivalent. At the time of writing there are no published results characterizing large families of Jordan pairs which are in fact strong Jordan pairs. In the rest of this section we provide a characterization of a family of weak Jordan pairs which are in fact Jordan pairs. This family includes all weak Jordan pairs likely to be of practical interest. The proof of the following technical lemma presents an interesting but quite involved mathematical argument. It may be skipped without any loss of understanding of what follows.

Lemma 5.4.7. *Let κ and λ be finitary spel-adjacencies in a digital space (V, π), such that κ is very tight. Let S be a finite $\kappa\lambda$-boundary in a binary picture (V, π, g). Then S is a $\kappa\lambda$-boundary in some finite picture (V, π, f).*

Proof. Let O be a κ-component of the set of 1-spels and Q be a λ-component of the set of 0-spels in (V, π, g) such that $S = \partial(O, Q)$. Since S is finite and nonempty, the same is true for $II(S)$ and $IE(S)$. We illustrate the argument presented in this proof by a sequence of figures, the first one of which is Figure 5.4.2, in which the underlying digital space is (Z^2, ω_2), κ is the very tight finitary spel-adjacency ω_2, and λ is the tight (but not very tight) spel-adjacency δ; see page 51.

Let O' be a finite κ-connected subset of O that contains $II(S)$, and let Q' be a finite λ-connected subset of Q that contains $IE(S)$. (Such O' and Q' exist, they can be constructed as indicated in the proof of Theorem 5.3.3.) Clearly, $S = \partial(O', Q')$. In Figure 5.4.3, O' and Q' are indicated by the heavier and lighter shades, respectively.

We define the sets $A = L_{\kappa \cup \lambda}^2(O' \cup Q')$ and $B = L_{\kappa \cup \lambda}^3(O' \cup Q') = L_{\kappa \cup \lambda}^1(A)$. (These are also indicated in Figure 5.4.3.) Since both κ and λ are finitary (and therefore so is $\kappa \cup \lambda$)

0	0	0	0	0	0	0	1	0	1	0	1	0	0	0	0	0	0
0	0	0	0	0	0	1	1	1	0	1	0	0	0	0	0	0	0
0	0	0	0	0	0	0	1	0	1	0	1	0	0	0	0	0	0
0	0	0	0	0	0	1	1	1	0	1	0	0	0	0	0	0	0
0	0	0	0	0	0	0	1	0	1	0	1	0	0	0	0	0	0
0	0	0	0	0	0	1	1	1	0	1	0	0	0	0	0	0	0
0	0	0	0	0	0	0	1	0	1	0	1	0	0	0	0	0	0
0	0	0	0	0	0	1	1	1	0	0	0	0	0	0	0	0	0
0	0	0	0	0	0	0	1	0	1	0	1	0	0	0	0	0	0
0	0	0	0	0	0	1	1	1	0	1	0	0	0	0	0	0	0
0	0	0	0	0	0	0	1	0	1	0	1	0	0	0	0	0	0
0	0	0	0	0	0	1	1	1	0	1	0	0	0	0	0	0	0
0	0	0	0	0	0	0	1	0	1	0	1	0	0	0	0	0	0
0	0	0	0	0	0	1	1	1	0	1	0	0	0	0	0	0	0
0	0	0	0	0	0	0	1	0	1	0	1	0	0	0	0	0	0

Figure 5.4.2. Illustration of the algorithm of Lemma 5.4.7. The relevant ω_2-component O of the set of 1-spels is indicated by the heavier shade, and the relevant δ-component Q of the set of 0-spels is indicated by the lighter shade. Both of these sets are infinite since the whole binary picture is defined by repetition of pairs of rows in the vertical direction. The finite $\omega_2\delta$-boundary S is indicated by the heavy line segments; the orientation is from the 1-spels to the 0-spels.

and $O' \cup Q'$ is finite, we know that both A and B are finite. We define the finite picture (V, π, f) by specifying f as follows. For any c in V,

$$f(c) \;=\; \begin{cases} g(c), & \text{if } c \in A \cup [(B - A) \cap R_\kappa(O)]\,, \\ 1, & \text{if } c \in (B - A) - R_\kappa(O)\,, \\ 0, & \text{if } c \notin B\,. \end{cases} \qquad (5.4.1)$$

The assignment of 1s and 0s by the function f is shown in Figure 5.4.4.

Let O'' be the κ-component of the set of 1-spels in (V, π, f) which contains O' and Q'' be the λ-component of the set of 0-spels in (V, π, f) which contains Q'. (These exist since O' is a subset of both O and A and so it is a κ-connected set of 1-spels in (V, π, f), and similarly Q' is a λ-connected set of 0-spels in (V, π, f). Both of these sets are illustrated in Figure 5.4.4.) Now we are going to show that $S = \partial(O'', Q'')$. This is sufficient to prove the theorem. Clearly, $S = \partial(O', Q')$ is a subset of $\partial(O'', Q'')$. To prove the converse, we show now that O'' is a subset of O and Q'' is a subset of Q.

Suppose that d is in O''. Let c be any element of $II(S)$. Then c is in O' (and hence in O'') and so there exists a κ-path $\langle c^{(0)}, \cdots, c^{(K)} \rangle$ in O'' from c to d. Now we use induction

0	0	0	0	0	0	0	1	0	1	0	1	0	0	0	0	0	0
0	0	0	0	0	0	1	1	1	0	1	0	0	0	0	0	0	0
0	0	0	0	0	0	0	1	0	1	0	1	0	0	0	0	0	0
0	0	0	0	0	0	1	1	1	0	1	0	0	0	0	0	0	0
0	0	0	0	0	0	0	1	0	1	0	1	0	0	0	0	0	0
0	0	0	0	0	0	1	1	1	0	1	0	0	0	0	0	0	0
0	0	0	0	0	0	0	1	0	1	0	1	0	0	0	0	0	0
0	0	0	0	0	0	1	1	1	0	0	0	0	0	0	0	0	0
0	0	0	0	0	0	0	1	0	1	0	1	0	0	0	0	0	0
0	0	0	0	0	0	1	1	1	0	1	0	0	0	0	0	0	0
0	0	0	0	0	0	0	1	0	1	0	1	0	0	0	0	0	0
0	0	0	0	0	0	1	1	1	0	1	0	0	0	0	0	0	0
0	0	0	0	0	0	0	1	0	1	0	1	0	0	0	0	0	0
0	0	0	0	0	0	1	1	1	0	1	0	0	0	0	0	0	0
0	0	0	0	0	0	0	1	0	1	0	1	0	0	0	0	0	0

Figure 5.4.3. Illustration of the algorithm of Lemma 5.4.7 (continued). The ω_2-connected set O' of 1-spels is indicated by the heavier shade, and the δ-connected set Q' of 0-spels is indicated by the lighter shade. Both of these sets are finite, and S, indicated by the heavy line segments, is the boundary between these finite sets. (The orientation is from the 1-spels to the 0-spels.) The double lines indicate the edges of the regions occupied by A and B, as defined in the proof of the lemma. The region $B - A$ is indicated with the very light shade.

to show that every element of this κ-path (in particular, d) is in O. Clearly, $c^{(0)}=c$ is in O. Now suppose that $c^{(k-1)}$ is in O. It follows that $c^{(k)} \in R_\kappa(O)$ and so (by (5.4.1) and using the fact that it is a 1-spel of (V, π, f)) we get that $c^{(k)}$ is a 1-spel of (V, π, g). Since O is a κ-component of the set of 1-spels in (V, π, g), it follows that $c^{(k)}$ is in O.

Suppose that d is in Q''. Let c be any element of $IE(S)$. Then c is in Q' (and hence in Q''), and so there exists a λ-path $\langle c^{(0)}, \cdots, c^{(K)} \rangle$ in Q'' from c to d. Now we use induction to show that every element of this λ-path (in particular, d) is in $Q \cap A$. Clearly, $c^{(0)}=c$ is in $Q \cap A$. Now suppose that $c^{(k-1)}$ is in $Q \cap A$. It follows that $c^{(k)} \in R_\lambda(A)$, which is a subset of B. So (by (5.4.1) and using the fact that it is a 0-spel of (V, π, f)) we get that $c^{(k)}$ is a 0-spel of (V, π, g). Since Q is a λ-component of the set of 0-spels in (V, π, g), it follows that $c^{(k)}$ is in Q. To complete the induction, all we have to do is to show that $c^{(k)}$ is in A. We know that $c^{(k)}$ is a 0-spel of (V, π, f) in $Q \cap B$. To finish the proof we show that all spels in $Q \cap (B - A)$ are 1-spels of (V, π, f). By (5.4.1), it is sufficient to show that $Q \cap R_\kappa(O) \subset A$. Suppose that $c \in Q \cap R_\kappa(O)$. Since κ was assumed to be very tight, there exists a π-path $\langle e^{(0)}, \cdots, e^{(T)} \rangle$ from an element $e^{(0)}$ of O to c such that $(e^{(0)}, e^{(t)}) \in \kappa$ for $1 \leq t \leq T$ (see (5.3.3)). Now consider the s, $0 \leq s < T$, such that $g(e^{(s)}) = 1$ and $g(e^{(t)}) = 0$ for $s < t \leq T$. (Since $g(e^{(0)}) =$

0	0	0	0	0	0	0	0	0	0	0	0	0	0	0	0	0	0
0	0	0	0	0	0	0	0	0	0	0	0	0	0	0	0	0	0
0	1	1	1	1	1	0	1	0	1	1	1	1	1	1	0	0	0
0	1	1	0	0	0	1	1	1	0	1	0	0	1	1	0	0	0
0	1	1	0	0	0	0	1	0	1	0	1	0	1	1	1	1	0
0	1	1	0	0	0	1	1	1	0	1	0	0	0	0	1	0	0
0	1	1	0	0	0	0	1	0	1	0	1	0	0	1	1	1	0
0	1	1	0	0	0	1	1	1	0	0	0	0	0	0	1	0	0
0	1	1	0	0	0	0	1	0	1	0	1	0	0	1	1	1	0
0	1	1	0	0	0	1	1	1	0	1	0	0	0	0	1	0	0
0	1	1	0	0	0	0	1	0	1	0	1	0	1	1	1	1	0
0	1	1	0	0	0	1	1	1	0	1	0	0	1	1	0	0	0
0	1	1	1	1	1	0	1	0	1	1	1	1	1	1	0	0	0
0	0	0	0	0	0	0	0	0	0	0	0	0	0	0	0	0	0
0	0	0	0	0	0	0	0	0	0	0	0	0	0	0	0	0	0

Figure 5.4.4. Illustration of the algorithm of Lemma 5.4.7 (continued). The 1s and 0s are assigned according to f as defined in (5.4.1). The ω_2-component O'' of the set of 1-spels is indicated by the heavier shade and the δ-component Q'' of the set of 0-spels is indicated by the lighter shade. Both of these sets are finite and S, indicated by the heavy line segments, is the boundary between these finite sets. (The orientation is from the 1-spels to the 0-spels.) The double lines indicate the edges of the regions occupied by A and B, as defined in the proof of the lemma.

1 and $g(e^{(T)}) = g(c) = 0$, such an s exists.) Since $e^{(0)}$ is in O and $(e^{(0)}, e^{(s)}) \in \kappa$, $e^{(s)}$ is in O. Since $e^{(T)} = c$ is in Q, $e^{(s+1)}$ is in Q. It follows that $e^{(s)}$ is in $II(S)$ and hence in O'. Therefore $e^{(0)} \in L^1_\kappa(O')$, and $c = e^{(T)} \in L^2_\kappa(O')$, which is a subset of A. □

Theorem 5.4.8. *Let $\{\kappa, \lambda\}$ be a weak Jordan pair for a digital space (V, π) such that κ and λ are both finitary and at least one of them is very tight. Then $\{\kappa, \lambda\}$ is a Jordan pair for (V, π).*

Proof. Without loss of generality, we may assume that κ is very tight. Let S be an arbitrary finite $\kappa\lambda$-boundary in an arbitrary binary picture (V, π, g). By Lemma 5.4.7, S is a $\kappa\lambda$-boundary in some finite picture (V, π, f). It follows that S is near-Jordan. □

To end this section on a philosophical note, we discuss what this theorem says to us: although it is perfectly reasonable to assume that in any practical application a binary picture will be a finite one, making this assumption may not help us much in deriving useful properties of binary pictures. It has been our general experience that, for the kind of spel-adjacencies one is likely to use in practice, the interesting properties of finite pictures also turn out to be properties of binary pictures in general. For this reason, finite pictures per se are not extensively discussed in this book.

5.5. New Jordan Pairs from Old Ones

In this section we prove a few additional results which help us to obtain new Jordan pairs from old ones. As opposed to Theorem 5.4.6 in which the digital space is kept fixed, the results below will allow us to find (strong) Jordan pairs in one digital space based on (strong) Jordan pairs in another digital space (but using the same set of spels). For this reason, for any spel-adjacency ρ in a digital space (V, π) and for any subsets O and Q of V, we introduce the notation

$$\partial_\rho(O, Q) = \{(c, d) \mid c \in O, \, d \in Q, \, (c, d) \in \rho\}. \tag{5.5.1}$$

Note that $\partial_\pi(O, Q)$ is just the boundary $\partial(O, Q)$ between O and Q. Recall also that (V, ρ) is a digital space and so $\partial_\rho(O, Q)$ is the boundary between O and Q in that space.

In the following proofs we repeatedly use the fact (see Exercise 3.2) that if S is a near-Jordan surface in a digital space (V, π), then for any π-path $\langle c^{(0)}, \cdots, c^{(K)} \rangle$ from an element of the interior of S to an element of the exterior of S there is a k, $1 \le k \le K$, such that $(c^{(k-1)}, c^{(k)}) \in S$.

Lemma 5.5.1. *Let ρ be a spel-adjacency in a digital space (V, π), let T be a near-Jordan surface in the digital space (V, ρ), and let $S = T \cap \pi$. Then S is a near-Jordan surface in the digital space (V, π).*

Proof. Let c be any element in the immediate interior of T in (V, ρ) and d be any element in the immediate exterior of T in (V, ρ). Since T is a surface, such c and d exist. Let $\langle c^{(0)}, \cdots, c^{(K)} \rangle$ be a π-path from c to d. Since $\pi \subset \rho$, $\langle c^{(0)}, \cdots, c^{(K)} \rangle$ is also a ρ-path and (since T is near-Jordan in (V, ρ)) we have for some k, $1 \le k \le K$, that $(c^{(k-1)}, c^{(k)}) \in T$. Consequently, $(c^{(k-1)}, c^{(k)}) \in S$, and S is nonempty. That it is near-Jordan follows (by the same argument) from the fact that the immediate interior of S in (V, π) is a subset of the immediate interior of T in (V, ρ) and the immediate exterior of S in (V, π) is a subset of the immediate exterior of T in (V, ρ). $\square$

As an example, consider the digital space (H, ε) based on the hexagonal grid H with edge-adjacency as proto-adjacency (see Figure 2.2.3). On the left of Figure 5.5.1 we represent a near-Jordan surface T in this digital space. Using the isomorphism i^{-1} from (H, ε) to (Z^2, β) (the inverse of the isomorphism i defined by (3.4.3)), we know (by Theorem 3.4.1) that $T' = \{ (i^{-1}(c), i^{-1}(d)) \mid (c, d) \in T \}$ is a near-Jordan surface in (Z^2, β). Then Lemma 5.5.1 implies that $S = T' \cap \omega_2$ is a near-Jordan surface in (Z^2, ω_2). This surface is represented on the right in Figure 5.5.1.

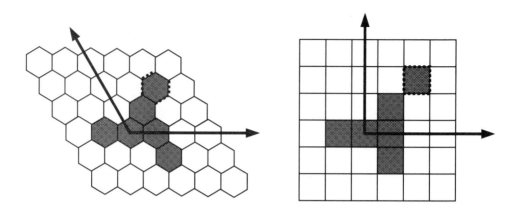

Figure 5.5.1. Demonstration of Lemma 5.5.1. On the left is a representation of the digital space (H, ε), in which the surface T formed by the ordered pairs of spels corresponding to shaded hexagons and edge-adjacent unshaded hexagons is near-Jordan. Let T' be the corresponding surface induced by the isomorphism from (H, ε) to the digital space (Z^2, β). On the right is a representation of the digital space (Z^2, ω_2), in which the surface $S = T' \cap \omega_2$ (formed by the ordered pairs of spels corresponding to shaded squares and edge-adjacent unshaded squares) is near-Jordan, as it should be according to Lemma 5.5.1. The four heavily drawn broken edges on the left represent surfels which correspond to the surfels represented by the four heavily drawn broken edges on the right. Note that the fifth edge of the hexagon on the left (which also represents a surfel in T) has no corresponding edge on the right because the corresponding surfel in T' is not in ω_2.

Lemma 5.5.2. *Let (V, π, f) be a binary picture over (V, π), and let κ, λ, and ρ be tight spel-adjacencies in (V, π) such that ρ is very tight and $\rho \subset \kappa \cap \lambda$. If O is a κ-component of the set of 1-spels and Q is a λ-component of the set of 0-spels such that $\partial_\pi(O, Q)$ is a near-Jordan surface in (V, π), then $\partial_\rho(O, Q)$ is a near-Jordan surface in (V, ρ).*

Proof. Let $S = \partial_\pi(O, Q)$ and $T = \partial_\rho(O, Q)$. Clearly, $S \subset T$. We adopt the convention that when we talk about the interior of S we mean its interior in (V, π) and when we talk about the interior of T we mean its interior in (V, ρ). Similar conventions apply to the exteriors, immediate interiors, and immediate exteriors.

Let $\langle c^{(0)}, \cdots, c^{(K)} \rangle$ be an arbitrary ρ-path from the immediate interior of T to the immediate exterior of T. We need to show that it crosses T. Since κ and λ are tight, it follows (by Theorem 5.3.6) that $O \subset I(S)$ and $Q \subset E(S)$. On the other hand, by definition, $II(T) \subset O$ and $IE(T) \subset Q$. Therefore, $c^{(0)} \in I(S)$ and $c^{(K)} \in E(S)$. In view of Lemma 3.2.1, there must exist a k, $1 \leq k \leq K$, such that $c^{(k-1)} \in I(S)$ and $c^{(k)} \in E(S)$. Our proof is complete if we can show that in fact $c^{(k-1)} \in O$ and $c^{(k)} \in Q$, since this implies that $(c^{(k-1)}, c^{(k)}) \in T$.

In this paragraph we show that $c^{(k-1)} \in O$. Since ρ is very tight, there exists a π-path $\langle e^{(0)}, \cdots, e^{(L)} \rangle$ from $c^{(k-1)}$ to $c^{(k)}$ such that $(c^{(k-1)}, e^{(l)}) \in \rho$ for $0 < l \leq L$. Since this is a π-path from the interior to the exterior of the surface S, which is near-Jordan in (V, π), there must be an l, $1 \leq l \leq L$, such that $(e^{(l-1)}, e^{(l)}) \in S$. If $l = 1$, then $c^{(k-1)} = e^{(l-1)}$ is in O, and we are done. Otherwise, if $c^{(k-1)}$ were a 0-spel, then it would be in Q (since it is ρ-adjacent and hence λ-adjacent to the spel $e^{(l)}$ in Q) and hence in $E(S)$. This would contradict Lemma 3.2.2. Therefore $c^{(k-1)}$ must be a 1-spel. Since it is ρ-adjacent and hence κ-adjacent to the spel $e^{(l-1)}$ in O, it is itself in O.

To show that $c^{(k)} \in Q$, essentially we repeat the same argument, using a π-path from $c^{(k)}$ to $c^{(k-1)}$ such that $c^{(k)}$ is ρ-adjacent to every other point on the path. $\square$

Note that the assumption that ρ is very tight is essential for this proof; the lemma would in fact be false without it. To see this, let $(V, \pi) = (Z, \nu_1)$, where we recall that $(c, d) \in \nu_1$ if, and only if, $|c - d| = 1$. Consider the binary picture over this digital space in which c is a 1-spel if, and only if, $|c| = 1$ or $|c| = 3$. Let κ, λ, and ρ all be the same spel-adjacency in this digital space, namely, ν_3 (see page 117). This is a tight spel-adjacency. Let $O = \{-1\}$ and $Q = \{0\}$. Then O is a ν_3-component of the set of 1-spels and Q is a ν_3-component of the set of 0-spels, such that $\partial_{\nu_1}(O, Q) = \partial_{\nu_3}(O, Q) = \{(-1, 0)\}$ is a near-Jordan surface in (Z, ν_1). So all the premises of the lemma are satisfied, except that $\rho = \nu_3$ is very tight (clearly, it is not), but the conclusion of the lemma (namely, that $\partial_{\nu_3}(O, Q)$ is a near-Jordan surface in (Z, ν_3)) is false, as can be seen by considering the ν_3-path $\langle -1, 2, 1, 0 \rangle$ from the immediate interior to the immediate exterior of $\partial_{\nu_3}(O, Q)$.

Theorem 5.5.3. *Let κ, λ, and ρ be tight spel-adjacencies in a digital space (V, π) such that $\rho \subset \kappa \cap \lambda$.*

(i) *If $\{\kappa, \lambda\}$ is a strong Jordan pair for (V, ρ), then it is a strong Jordan pair for (V, π).*

(ii) *If ρ is very tight and $\{\kappa, \lambda\}$ is a strong Jordan pair (respectively, a Jordan pair) for (V, π), then it is a strong Jordan pair (respectively, a Jordan pair) for (V, ρ).*

Proof. Let (V, π, f) be a binary picture over the digital space (V, π), O be a κ-component of the set of 1-spels, and Q be a λ-component of the set of 0-spels. Let $S = \partial_\pi(O, Q)$ and $T = \partial_\rho(O, Q)$. Clearly, $S = T \cap \pi$.

To prove (i), assume that $\{\kappa, \lambda\}$ is a strong Jordan pair for (V, ρ). We need to show that if an S as defined above is a surface (i.e., if it is not empty), then it is near-Jordan in (V, π). Since the nonemptiness of S implies that of T, by our assumption T is a near-Jordan surface in (V, ρ). By Lemma 5.5.1, it follows that S is a near-Jordan surface in (V, π).

To prove (ii), assume that $\{\kappa, \lambda\}$ is a strong Jordan pair (respectively, a Jordan pair) for (V, π). We need to show that if a T as defined above is a surface (respectively, a finite surface), then it is near-Jordan in (V, ρ). In the next paragraph, we show that if T is a surface, then so is S. It is clearly finite if T is finite. By our assumption, S is a near-Jordan surface in (V, π), and so, by the previous lemma, T is a near-Jordan surface in (V, ρ).

Using the fact that ρ is very tight, now we show that if T is nonempty, then S is also nonempty. Suppose that $(c, d) \in T$. Then $(c, d) \in \rho$, and so there exists a π-path $\langle c^{(0)}, \cdots, c^{(K)} \rangle$ from c to d such that $(c^{(0)}, c^{(k)}) \in \rho$, for $1 \leq k \leq K$. Let $c^{(k-1)}$ be the last 1-spel on this path; it exists since c is a 1-spel and d is a 0-spel. Then $c^{(k-1)}$ must be in O because $c^{(0)} \in O$ is ρ-adjacent to it, and $c^{(k)}$ must be in Q because $\langle c^{(k)}, \cdots, c^{(K)} \rangle$ is a ρ-path of 0s and $c^{(K)}$ is in Q. This shows that $(c^{(k-1)}, c^{(k)}) \in S$. $\square$

We postpone illustrating this theorem until the next chapter; it will be more interesting to do so after we have found some useful examples of (strong) Jordan pairs (see, however, Exercise 5.19). We complete this chapter by discussing the production of new Jordan pairs from old ones by using isomorphisms. For this we need the following concept. If i is an isomorphism from a digital space (V, π) to a digital space (V', π') and ρ is a spel-adjacency in (V, π), then we call the spel-adjacency ρ' in (V', π') the *isomorphic image* of ρ if it satisfies (3.5.5).

Theorem 5.5.4. *Suppose that there is an isomorphism i from the digital space (V, π) to the digital space (V', π'), κ and λ are spel-adjacencies in (V, π), and κ' and λ' are their respective isomorphic images in (V', π'). If $\{\kappa, \lambda\}$ is a strong Jordan pair (respectively, a Jordan pair or a weak Jordan pair) for (V, π), then $\{\kappa', \lambda'\}$ is a strong Jordan pair (respectively, a Jordan pair or a weak Jordan pair) for (V', π').*

Proof. First we note that if κ and λ are tight, then so are κ' and λ' (Exercise 5.9).

Let (V', π', f') be an arbitrary binary picture over (V', π') and define the function f on V by $f(c) = f'(i(c))$. Then (V, π, f) is a binary picture over (V, π). Furthermore, if O' is a κ'-component of the set of the set of 1-spels in (V', π', f'), then $O = \{c \mid i(c) \in O'\}$ is κ-component of the set of 1-spels in (V, π, f). (Indeed, O consists of 1-spels only and κ-paths $\langle c^{(0)}, \cdots, c^{(K)} \rangle$ in O correspond to κ'-paths $\langle i(c^{(0)}), \cdots, i(c^{(K)}) \rangle$ in O', since κ' is the isomorphic image of κ.) Similarly, if Q' is a λ'-component of the set of 0-spels in (V', π', f'), then $Q = \{d \mid i(d) \in Q'\}$ is λ-component of the set of 0-spels in (V, π, f). Given a $\kappa'\lambda'$-boundary $S' = \partial_{\pi'}(O', Q')$ in (V', π', f'), let us define $S = \partial_\pi(O, Q)$. Then, by (5.5.1) and (3.4.1),

$$
\begin{aligned}
(c, d) \in S &\Leftrightarrow c \in O \text{ and } d \in Q \text{ and } (c, d) \in \pi \\
&\Leftrightarrow i(c) \in O' \text{ and } i(d) \in Q' \text{ and } (i(c), i(d)) \in \pi' \qquad (5.5.2) \\
&\Leftrightarrow (i(c), i(d)) \in S' \, .
\end{aligned}
$$

This implies that S is a surface (since it is nonempty) and that (according to Theorem 3.4.1) if S is near-Jordan in (V, π), then S' is near-Jordan in (V', π').

So, if $\{\kappa, \lambda\}$ is a strong Jordan pair for (V, π), then every $\kappa'\lambda'$-boundary S' in every binary picture (V', π', f') over (V', π') is near-Jordan in (V', π'), since the corresponding $\kappa\lambda$-boundary S in (V, π, f) is near-Jordan in (V, π). The rest of the proof follows similarly, by observing that if S' is finite, then so is S and that if (V', π', f') is a finite picture, then so is (V, π, f). □

5.6. Exercises

5.1. Let M be any set, ρ be a binary relation on M, and let A be any subset of M. Prove that the following three conditions are equivalent:

(i) A is a ρ-component of M;
(ii) for some positive integer l, $L_\rho^l(A) = A$;
(iii) for all positive integers l, $L_\rho^l(A) = A$.

5.2. Let (V, π, f) be the binary picture in which V consists of all the faces of the sugar cubes in Figure 1.9.1, π is defined as in Figure 2.2.4, and $f(c) = 1$ if, and only if, $c \in \{I1, I2, I3, I5, I6, I15, III2, III3, III4, III5, III6, IV2\}$.

(i) Let O be the π-component of the set of 1-spels containing $I1$ and Q be the π-component of the set of 0-spels containing $III1$. Is $\partial(O, Q)$ a Jordan surface?

(ii) Define a fuzzy spel affinity on (V, π) by

$$\psi(c, d) = \begin{cases} 0, & \text{if } (c, d) \notin \pi, \\ 1 - |f(c) - f(d)|, & \text{otherwise.} \end{cases} \tag{5.6.1}$$

Let

$$f'(c) = \begin{cases} 1, & \text{if } \mu_\psi(I1, c) \geq 0.5, \\ 0, & \text{otherwise.} \end{cases} \tag{5.6.2}$$

Let O' be the π-component of the set of 1-spels of (V, π, f') containing $I1$ and Q' be the π-component of the set of 0-spels (V, π, f') containing III. Is $\partial(O', Q')$ a Jordan surface?

5.3. Modify the Dynamic Program for Fuzzy Objects so that it can also be used to produce a maximum strength π-path of the kind illustrated in Figure 5.2.3.

5.4. Prove that if ρ is a local spel-adjacency in (Z^2, ω_2), then ρ is 2-limited and that if ρ is a local spel-adjacency in (Z^3, ω_3), then ρ is 3-limited.

5.5. Prove that if $N \geq 4$, then there is a local spel-adjacency in (Z^N, ω_N) which is not tight.

5.6. Show that the following is *false*. "Let N be a positive integer and ρ be a local and tight spel-adjacency in (Z^N, ω_N). For every pair of ρ-adjacent spels c and d in (Z^N, ω_N), there is a ρ-tight ω_N-path $\langle c^{(0)}, \cdots, c^{(K)} \rangle$ from c to d such that, for $1 \leq k < K$, the Voronoi neighborhood of $c^{(k)}$ contains the intersection of the Voronoi neighborhoods of c and d." (*Hint:* let ρ be the local spel-adjacency in (Z^5, ω_5) for which $(c, d) \in \rho$ if, and only if, either $c_5 \neq d_5$ or $\sum_{n=1}^{4} |c_n - d_n| \neq 2$.)

5.7. Let ρ be a spel-adjacency in a digital space (V, π). Show that
  (i) ρ is very tight if, and only if, $L_\rho^1(\{c\})$ is π-connected for all spels c;
  (ii) if ρ is very tight, then ρ is tight;
  (iii) if ρ is 2-limited, then ρ is very tight.

5.8. Show that ν is a tight, but not a very tight spel-adjacency in (Z^2, χ).

5.9. Show that if i is an isomorphism from a digital space (V, π) to a digital space (V', π'), ρ is a spel-adjacency in (V, π), and ρ' is its isomorphic image in (V', π'), as defined by (3.5.5), then ρ is a tight (respectively, very tight) spel-adjacency in (V, π) if, and only if, ρ' is a tight (respectively, very tight) spel-adjacency in (V', π').

5.10. Let ρ be an arbitrary spel-adjacency in (Z^4, ω_4) and let ϕ be as defined on page 42. Is it true that if S is a finite $\phi\rho$-boundary in a binary picture over (Z^4, ω_4), then S is a $\phi\rho$-boundary in some finite binary picture over (Z^4, ω_4)?

	0	0	0	0	0	0	0	0	0	0	0	0	0	0	0	0	0	
.	0	0	0	0	0	0	0	0	1	0	0	0	0	0	0	0	0	.
.	0	0	0	0	0	0	0	1	0	1	0	0	0	0	0	0	0	.
.	0	0	0	0	1	0	1	0	1	0	1	0	1	0	0	0	0	.
.	0	0	0	0	0	0	0	1	0	1	0	0	0	0	0	0	0	.
.	0	0	0	0	0	0	0	0	1	0	0	0	0	0	0	0	0	.
.	0	0	0	0	0	0	0	0	0	0	0	0	0	0	0	0	0	.

Figure 5.6.1. Illustration of the necessity that both κ and λ be tight in Theorem 5.3.7.

5.11. Define a spel-adjacency ρ in $\left(Z^2, \omega_2\right)$ such that in the binary picture depicted in Figure 5.6.1 has a $\rho\omega_2$-boundary whose not ρ-connected interior consists of the two pixels shaded in Figure 5.6.1.

5.12. Give an example of a finite picture (Z, ν_1, f) in which there is a $\nu_4\nu_6$-boundary whose interior is not ν_4-connected.

5.13. Prove that if O is a χ-connected subset of Z^2 and, for $i = 1, 2$, $c_i \in O$, $d_i \in \overline{O}$, $(c_i, d_i) \in \chi$ with $d_1 \neq d_2$, then d_1 is not χ-connected in $\overline{O}$ to d_2.

5.14. Looking at Figure 5.4.1, find all possible combinations of a ν-component O of the set of 1-spels and a ν-component Q of the set of 0-spels. Check individually for each combination that $\partial(O, Q)$ is either empty or is a near-Jordan surface in the digital space $\left(Z^2, \chi\right)$, but is not a near-Jordan surface in the digital space $\left(Z^2, \omega_2\right)$.

5.15. For any positive integer N, show that if κ and λ are very tight local spel-adjacencies on $\left(Z^N, \omega_N\right)$, then $\{\kappa, \lambda\}$ is a Jordan pair for $\left(Z^N, \omega_N\right)$ if, and only if, every $\kappa\lambda$-boundary in every finite picture over $\left(Z^N, \omega_N\right)$ is near-Jordan.

5.16. For any positive integer N, show that $\{\omega_N, \omega_N\}$ is a strong Jordan pair (equivalently a Jordan pair or a weak Jordan pair) for $\left(Z^N, \omega_N\right)$ if, and only if, $N = 1$.

5.17. Show that $\{\beta, \gamma\}$ (see Figure 2.2.1) is not a weak Jordan pair for $\left(Z^2, \omega_2\right)$.

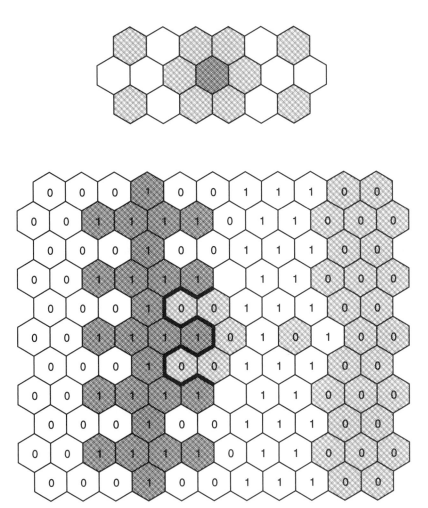

Figure 5.6.2. On top we indicate an adjacency in the digital space (H, ε): the lighter shaded pixels are adjacent to the darker shaded pixel. (This adjacency is an analog in the hexagonal grid of the δ of Figure 2.2.1 in the square grid.) Below it we try to produce an analog in the hexagonal grid of the binary picture of Figure 5.4.2. Note that the values assigned to two pixels are unspecified. Irrespective of the values which may be assigned to these pixels, the darker shaded pixels form an ε-connected subset in the set of 1-spels and the set of lighter shaded pixels is connected (according to the adjacency specified on top) in the set of 0-spels. The heavy lines indicate the boundary between these two sets.

5.18. Show that for any assignment of 1s and 0s to the two pixels whose value is not specified in Figure 5.6.2, the boundary between the ε-component of the set of 1-spels which contains the shaded ones and the component (according to the adjacency specified on the top of Figure 5.6.2) of the set of 0-spels which contains the shaded ones is near-Jordan. (Contrast this with the boundary in Figure 5.4.2. Just as in that figure, we assume that the unspecified pixels to the left and to the right of the specified ones are 0-spels and that the unspecified pixels above and below the specified ones are given values according to infinite repeats of the top two and bottom two specified rows.)

5.19.   Using Theorem 5.5.3 and the material on page 123, show that, for the binary relation on Z^2 defined by

$$\mu = \left\{ \, ((c_1, c_2), (d_1, d_2)) \mid [\, c_2 = d_2 \ \& \ c_1 \times d_1 \geq 0\,] \text{ or} \right.$$
$$\left. [\, |c_2 - d_2| = 1 \ \& \ c_1 = d_1 = 0 \,] \, \right\} , \qquad (5.6.3)$$

$\{\nu, \nu\}$ is a strong Jordan pair for (Z^2, μ).

6
Simply Connected Digital Spaces

"Only connect the prose and the passion, and both will be exulted, and human love will be seen at its height."

E. M. Forster, *Howards End*, Chapter XXII.

6.1. N-Simply Connected Digital Spaces

Let S be a surface in a digital space (V, π). For any practical application, it would be impossible to determine whether S is near-Jordan by examining all π-paths from $II(S)$ to $IE(S)$. It is desirable to have a result which says that S is near-Jordan if some local condition is satisfied at every surfel of S. We illustrate this with the digital space (Z^3, ω_3). Let (c, d) be a surfel of a surface S in (Z^3, ω_3). If one of the edges of (c, d) is "loose" (in the sense that no other surfel in the surface shares this edge; see Figure 6.1.1), then one is able to get from c to d via an ω_3-path of length 3 which does not cross S. For S to be near-Jordan, it is

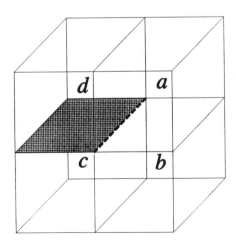

Figure 6.1.1. A surfel (c, d) of a surface S in (Z^3, ω_3) has a "loose" edge, indicated by the heavy broken line. The ω_3-path $\langle c, b, a, d \rangle$ from c to d does not cross S, implying that S is not near-Jordan.

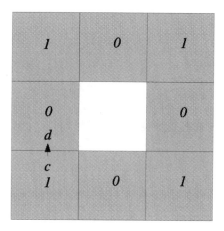

Figure 6.1.2. A binary picture over digital space (V, π) which is not "simply connected." V is the set of the eight shaded squares, and two elements in V are proto-adjacent if, and only if, they share an edge. In the binary picture the 1-spels are labeled *1* and the 0-spels are labeled *0*. The $\pi\pi$-boundary $\{(c, d)\}$ between the π-component $\{c\}$ of 1-spels and the π-component $\{d\}$ of 0-spels appears to be near-Jordan locally (and in fact is 2-locally-Jordan according to the definition given in the next section) but is not near-Jordan in the digital space (V, π).

in particular necessary that ω_3-paths of length not more than 3 from c to d must cross S. It would be very useful if this local condition were also sufficient. However, this is not the case for an arbitrary digital space (V, π), even if we restrict our attention to finite $\pi\pi$-boundaries in binary pictures over (V, π).

We demonstrate this with an example. In Figure 6.1.2 we show a digital space (V, π), where V is the set of shaded squares and two spels are proto-adjacent if, and only if, they share an edge. The 1s and 0s of Figure 6.1.2 define a binary picture over (V, π). Let $S = \partial(O, Q)$, where $O = \{c\}$ is a π-component of 1-spels and $Q = \{d\}$ is a π-component of 0-spels. S, whose only element is the surfel represented by the edge with an arrow, satisfies the property discussed in the last paragraph, namely, that every π-path of length not more than 3 from c to d must cross S. However, S is not near-Jordan, since there is a π-path (of length 7) from c to d which does not cross S. Consideration of the difference between the situations in Figures 6.1.1 and 6.1.2 led us to introduce the concept of a simply connected digital space.

In classical topology, a simply connected space is (intuitively speaking) a connected space in which every loop can be continuously pulled to a point without leaving the space [31]. There is an infinity of corresponding notions for digital spaces. For every positive integer N (reflecting how large a digital step is allowed to replace the notion of continuity), there is a class of N-simply connected digital spaces, whose definition now follows.

If

$$P = \langle c^{(1)}, \cdots, c^{(m)}, d^{(0)}, \cdots, d^{(n)}, e^{(1)}, \cdots, e^{(l)} \rangle \qquad (6.1.1)$$

and

$$P' = \langle c^{(1)}, \cdots, c^{(m)}, f^{(0)}, \cdots, f^{(k)}, e^{(1)}, \cdots, e^{(l)} \rangle \qquad (6.1.2)$$

are π-paths in the digital space (V, π) such that

$$f^{(0)} = d^{(0)}, \quad f^{(k)} = d^{(n)}, \quad \text{and} \quad 1 \le k + n \le N + 2, \qquad (6.1.3)$$

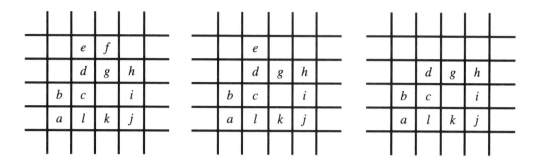

Figure 6.1.3. Demonstration of equivalence of π-paths in (Z^2, ω_2): $\langle a, b, c, d, e, f, g, h, i, j, k, l, a \rangle$ is 2-equivalent to $\langle a, b, c, d, g, h, i, j, k, l, a \rangle$, since it is elementarily 2-equivalent to $\langle a, b, c, d, e, d, g, h, i, j, k, l, a \rangle$, which is elementarily 2-equivalent to $\langle a, b, c, d, g, h, i, j, k, l, a \rangle$.

then P and P' are said to be *elementarily N-equivalent*. (Note that in this definition, m or l or both in (6.1.1) may be zero. In other words, the difference between elementarily N-equivalent π-paths may be at their "head" or at their "tail.") Two π-paths, P and P' in a digital space (V, π) are said to be *N-equivalent* if there is a sequence of π-paths $P_0, \cdots, P_L$ ($L \geq 0$) in the digital space such that $P_0 = P$, $P_L = P'$ and, for $1 \leq l \leq L$, P_{l-1} and P_l are elementarily N-equivalent.

We demonstrate the notion of 2-equivalent π-paths in (Z^2, ω_2) in Figure 6.1.3. That $\langle a, b, c, d, e, f, g, h, i, j, k, l, a \rangle$ is elementarily 2-equivalent to $\langle a, b, c, d, e, d, g, h, i, j, k, l, a \rangle$ follows by substituting in (6.1.1) $n = 2$, $d^{(0)} = e$, $d^{(1)} = f$, $d^{(2)} = g$ and in (6.1.2) $k = 2$, $f^{(0)} = e$, $f^{(1)} = d$, $f^{(2)} = g$. That $\langle a, b, c, d, e, d, g, h, i, j, k, l, a \rangle$ is elementarily 2-equivalent to $\langle a, b, c, d, g, h, i, j, k, l, a \rangle$ follows by substituting in (6.1.1) $n = 2$, $d^{(0)} = d$, $d^{(1)} = e$, $d^{(2)} = d$ and in (6.1.2) $k = 0$, $f^{(0)} = d$.

A *loop* (*of length K*) in a digital space (V, π) is a π-path $\langle c^{(0)}, \cdots, c^{(K)} \rangle$ such that $c^{(K)} = c^{(0)}$. (Compare this with the corresponding definition for loops in digraphs.) In particular, for any spel c, $\langle c \rangle$ is a loop and is called a *trivial loop*. (The notion of a trivial loop and the previously defined notion of a trivial path are identical; we select the terminology to be used depending on the context.) We note that any loop of length 1, 2, or 3 is automatically N-equivalent to a trivial loop for any positive integer N (Exercise 6.1). A digital space is said to be *N-simply connected* if every loop in the digital space is N-equivalent to a trivial loop.

As a simple example, we show that the digital space (V, π) of Figure 6.1.2 is 6-simply connected. If not, then there must be a loop of length greater than 0 which is not 6-equivalent to a trivial loop. Among all such loops choose one, $\langle c^{(0)}, \cdots, c^{(K)} \rangle$, of minimal length (i.e., any loop of length less than K is 6-equivalent to a trivial loop). Since $c^{(K)} = c^{(0)}$ and V has only 8 elements, there must be integers i and j such that $0 \leq i < j \leq K$, $j - i \leq 8$, and $c^{(j)} = c^{(i)}$. It follows that $\langle c^{(0)}, \cdots, c^{(K)} \rangle$ is elementarily 6-equivalent to the loop $\langle c^{(0)}, \cdots, c^{(i-1)}, c^{(i)} = c^{(j)}, c^{(j+1)}, \cdots, c^{(K)} \rangle$ of length $K - (j - i)$. Since the length of this loop is less than K, it is 6-equivalent to a trivial loop. That implies that $\langle c^{(0)}, \cdots, c^{(K)} \rangle$ itself is 6-equivalent to a trivial loop. This contradiction shows that (V, π) of Figure 6.1.2 is 6-simply connected. The proof just given can be generalized to arbitrary digital spaces (V, π) in which V has finitely many elements; see Exercise 6.3.

The reasons for studying N-simply connected digital spaces are the following. As we will see, many of the interesting digital spaces are N-simply connected for some small N. Using this fact, we will be able to provide local tests for near-Jordanness of surfaces in such digital spaces. Furthermore, this will lead to our being able to prove that certain pairs of spel-adjacencies are strong Jordan pairs and, consequently, that certain claims stated in Chapters 1 and 2 are in fact valid. Furthermore, it will be seen in Chapter 8 that simple connectedness will be found very useful for proving the correctness of boundary tracking algorithms.

We complete this section by discussing a result which helps us to obtain new N-simply connected digital spaces from existing ones.

Theorem 6.1.1. *If (V, π) is an N-simply connected digital space and ρ is an l-limited spel-adjacency in it, then (V, ρ) is an M-simply connected digital space with*

$$M = max(N, l - 1). \tag{6.1.4}$$

Proof. Clearly, (V, ρ) is a digital space. To show that it is M-simply connected, consider any loop $\langle c^{(0)}, \cdots, c^{(K)} \rangle$ in (V, ρ). Since ρ is an l-limited spel-adjacency in (V, π), for each $1 \leq k \leq K$ there exists a π-path $\langle c^{(k-1)} = c_k^{(0)}, \cdots, c_k^{(m_k)} = c^{(k)} \rangle$ in (V, π), and therefore in (V, ρ), of length $m_k \leq l$. Observing (6.1.1), (6.1.2), and (6.1.3), we see that the loops $\langle c^{(0)}, c^{(1)}, \cdots, c^{(K)} \rangle$ and $\langle c^{(0)} = c_1^{(0)}, \cdots, c_1^{(m_1)} = c^{(1)} = c_2^{(0)}, \cdots, c_2^{(m_2)}, \cdots \cdots \cdots, c_K^{(0)}, \cdots, c_K^{(m_K)} = c^{(K)} \rangle$ are $(l-1)$-equivalent, and hence M-equivalent, in (V, ρ). The second of these loops is also a loop in (V, π) and so, by the N-simple connectedness of (V, π), it is N-equivalent (and hence M-equivalent) in (V, π) (and hence in (V, ρ)) to a trivial loop. $\square$

6.2. Locally-Jordan Surfaces

Let S be a surface in a digital space (V, π). We say that a surfel (c, d) in (V, π) *crosses* S if exactly one of $(c, d) \in S$ or $(d, c) \in S$ is true. Let $P = \langle c^{(0)}, \cdots, c^{(K)} \rangle$ be a π-path in (V, π). We say that the *crossing parity $p_S P$ of P through S* is *even* (or zero, i.e., $p_S P = 0$) if the number of surfels among $(c^{(0)}, c^{(1)}), \cdots, (c^{(K-1)}, c^{(K)})$ that cross S is even and we say that it is *odd* (or one, i.e., $p_S P = 1$) if this number is odd. We use the notation $\oplus$ for *modulo 2 addition* of parities (i.e., $0 \oplus 0 = 1 \oplus 1 = 0$ and $0 \oplus 1 = 1 \oplus 0 = 1$).

It is easy to see (Exercise 6.8) that, for any surface S in a digital space, the crossing parity through S is even for any loop in the digital space whose length is not greater than two. Also, cyclic permutation of a loop does not influence its crossing parity through a surface S, since, for $1 \leq k \leq K$,

$$p_S \langle c^{(0)}, \cdots, c^{(k-1)}, c^{(k)}, \cdots, c^{(K)} \rangle = p_S \langle c^{(k)}, \cdots, c^{(K)} = c^{(0)}, \cdots, c^{(k-1)}, c^{(k)} \rangle . \tag{6.2.1}$$

In addition, reversing a π-path does not change its crossing parity. It is also easy to see that if $\langle c^{(0)}, \cdots, c^{(K)} \rangle$ and $\langle d^{(0)}, \cdots, d^{(L)} \rangle$ are π-paths such that $c^{(K)} = d^{(0)}$, then

$$p_S \langle c^{(0)}, \cdots, c^{(K)}, d^{(1)}, \cdots, d^{(L)} \rangle = p_S \langle c^{(0)}, \cdots, c^{(K)} \rangle \oplus p_S \langle d^{(0)}, \cdots, d^{(L)} \rangle . \tag{6.2.2}$$

Theorem 6.2.1. *If S is a near-Jordan surface in a digital space, then the crossing parity through S is odd for any π-path $P = \langle c^{(0)}, \cdots, c^{(K)} \rangle$ such that $(c^{(0)}, c^{(K)}) \in S$.*

Proof. First note that (according to Lemma 3.2.2), since S is near-Jordan, $I(S) \cap E(S) = \emptyset$ and $S = \partial(I(S), E(S))$. We prove by induction that, for any $0 \leq k \leq K$,

$$p_S \langle c^{(0)}, \cdots, c^{(k)} \rangle = \begin{array}{ll} 0, & \text{if } c^{(k)} \in I(S), \\ 1, & \text{otherwise.} \end{array} \tag{6.2.3}$$

Since $c^{(K)} \in E(S)$, this is sufficient to prove the theorem.

Clearly, (6.2.3) is true for $k = 0$. Suppose that (6.2.3) is true for some $k - 1$, where $1 \leq k \leq K$. We prove that it is also true for k. We repeatedly use the following special case of (6.2.2):

$$p_S \langle c^{(0)}, \cdots, c^{(k-1)}, c^{(k)} \rangle = p_S \langle c^{(0)}, \cdots, c^{(k-1)} \rangle \oplus p_S \langle c^{(k-1)}, c^{(k)} \rangle . \tag{6.2.4}$$

In case $c^{(k-1)} \in I(S)$, the first term on the right-hand side of (6.2.4) is 0, and the second term is 0 if $c^{(k)} \in I(S)$ and 1 otherwise. In case $c^{(k-1)} \notin I(S)$, the first term on the right-hand side of (6.2.4) is 1, and the second term is 1 if $c^{(k)} \in I(S)$ and 0 otherwise. In either case, (6.2.3) is true for k. $\square$

A surface S in a digital space (V, π) is said to be *N-locally-Jordan* (where N is a positive integer) if $p_S \langle c^{(0)}, \cdots, c^{(K)} \rangle$ is odd for any π-path such that $(c^{(0)}, c^{(K)}) \in S$ and $2 \leq K \leq N + 1$. By Theorem 6.2.1, if a surface S in a digital space (V, π) is near-Jordan, then it is *N-locally-Jordan* for all positive N. Much of what follows will lead to a converse of this statement for tight spel-adjacencies in *N*-simply connected digital spaces (Theorem 6.2.6).

Lemma 6.2.2. *Any loop of length not more than $N + 2$ has even crossing parity through any N-locally-Jordan surface in any digital space (V, π).*

Proof. We have already pointed out that the crossing parity through any surface is even for any loop of length not more than two.

Let S be an *N-locally-Jordan* surface. Consider a loop $L = \langle c^{(0)}, \cdots, c^{(K)} \rangle$ with $3 \leq K \leq N + 2$. If there does not exist a k, $1 \leq k \leq K$, such that $(c^{(k-1)}, c^{(k)})$ crosses S, then we are done. Otherwise, for such a k, exactly one of $(c^{(k-1)}, c^{(k)}) \in S$ or $(c^{(k)}, c^{(k-1)}) \in S$ is true. If $(c^{(k)}, c^{(k-1)}) \in S$, $P = \langle c^{(k)}, \cdots, c^{(K)} = c^{(0)}, \cdots, c^{(k-1)} \rangle$ is a π-path of length $K - 1$ with $2 \leq K - 1 \leq N + 1$, and so the fact that S is *N-locally-Jordan* implies $p_S P = 1$. By application of (6.2.1) and (6.2.2),

$$p_S L = p_S P \oplus p_S \langle c^{(k-1)}, c^{(k)} \rangle = 1 \oplus 1 = 0 . \tag{6.2.5}$$

If $(c^{(k-1)}, c^{(k)}) \in S$, a similar argument, which also makes use of the fact that reversing a π-path does not change its crossing parity, can be used to derive the same conclusion. $\square$

Lemma 6.2.3. *Let S be an N-locally-Jordan surface in a digital space (V, π). If P and P' are N-equivalent π-paths, then they have the same crossing parity through S.*

Proof. By the definition of *N-equivalent*, it is sufficient to prove that if P and P' satisfy (6.1.1), (6.1.2), and (6.1.3), then they have the same crossing parity through S.

By applying (6.2.2), we get

$$p_S P = p_S \langle c^{(1)}, \cdots, c^{(m)}, d^{(0)} \rangle \oplus p_S \langle d^{(0)}, \cdots, d^{(n)} \rangle \oplus p_S \langle d^{(n)}, e^{(1)}, \cdots, e^{(l)} \rangle \quad (6.2.6)$$

and

$$p_S P' = p_S \langle c^{(1)}, \cdots, c^{(m)}, f^{(0)} \rangle \oplus p_S \langle f^{(0)}, \cdots, f^{(k)} \rangle \oplus p_S \langle f^{(k)}, e^{(1)}, \cdots, e^{(l)} \rangle . \quad (6.2.7)$$

Therefore, using (6.1.3), the invariance of crossing parity under reversal, and (6.2.2), we get

$$
\begin{aligned}
p_S P \oplus p_S P' &= p_S \langle d^{(0)}, \cdots, d^{(n)} \rangle \oplus p_S \langle f^{(0)}, \cdots, f^{(k)} \rangle \\
&= p_S \langle d^{(0)}, \cdots, d^{(n)} = f^{(k)}, \cdots, f^{(0)} = d^{(0)} \rangle = 0 .
\end{aligned}
\quad (6.2.8)
$$

The last equality follows from the previous lemma combined with (6.1.3). $\square$

Up to now the results apply to digital spaces which do not have to be N-simply connected for any N. The next lemma makes essential use of N-simple connectedness.

Lemma 6.2.4. *If S is an N-locally-Jordan antisymmetric surface in an N-simply connected digital space (V, π), then S is near-Jordan if either (and hence both) of the following two equivalent conditions holds.*
   (i) *For any $c \in II(S)$ and $d \in II(S)$, there exists a π-path P from c to d such that $p_S P = 0$.*
   (ii) *For any $c \in IE(S)$ and $d \in IE(S)$, there exists a π-path P from c to d such that $p_S P = 0$.*

Proof. Evidently the two conditions are equivalent. Indeed, if $c \in II(S)$ and $d \in II(S)$, then there exist $c' \in IE(S)$ and $d' \in IE(S)$ such that $(c, c') \in S$ and $(d, d') \in S$. Hence if there exists a π-path of even crossing parity from c' to d', then there also exists one from c to d, so that (ii) implies (i). Similarly, (i) implies (ii).

In what follows, we prove that S is near-Jordan if (ii) holds. We do this by supposing that S is not near-Jordan and showing that this, together with (ii), leads to a contradiction.

First, we show that there exists a π-path $P_1 = \langle c^{(1)}, \cdots, c^{(K)} \rangle$ such that $c^{(1)} \in II(S)$, $c^{(K)} \in IE(S)$, and $p_S P_1 = 0$. Indeed, since S is not supposed to be near-Jordan, there is a π-path from $II(S)$ to $IE(S)$ that does not cross S. Clearly, any such π-path has the required properties.

Next, we show that there exists a π-path $P_3 = \langle e^{(1)}, \cdots, e^{(L)} \rangle$ such that $(e^{(1)}, e^{(L)}) \in S$ and $p_S P_3 = 0$. Let P_1 be the π-path of the last paragraph. Let $c^{(0)}$ be such that $(c^{(1)}, c^{(0)}) \in S$. By (ii), there exists a π-path $P_2 = \langle c^{(K)} = d^{(0)}, \cdots, d^{(L)} = c^{(0)} \rangle$ from $c^{(K)}$ to $c^{(0)}$ such that $p_S P_2 = 0$. Then $P_3 = \langle c^{(1)}, \cdots, c^{(K)}, d^{(1)}, \cdots, d^{(L)} \rangle$ is a π-path from $c^{(1)}$ to $d^{(L)}$ such that $(c^{(1)}, d^{(L)}) \in S$ and, by (6.2.2), $p_S P_3 = p_S P_1 \oplus p_S P_2 = 0$.

For a π-path P_3 satisfying the properties listed at the beginning of the previous paragraph, let $e^{(0)} = e^{(L)}$. By the antisymmetry of S, $p_S \langle e^{(0)}, e^{(1)} \rangle = 1$ and so, by (6.2.2), $P_4 = \langle e^{(0)}, e^{(1)}, \cdots, e^{(L)} \rangle$ is a loop such that $p_S P_4 = p_S \langle e^{(0)}, e^{(1)} \rangle \oplus p_S P_3 = 1$. Since (V, π) is N-simply connected, P_4 is N-equivalent to a trivial loop, whose crossing parity through S is zero. Since S is N-locally-Jordan, according to Lemma 6.2.3, we also have that $p_S P_4 = 0$, contradicting the fact that $p_S P_4 = 1$. $\square$

Interestingly, since the surface of Figure 6.1.2 has only one element, (i) and (ii) of Lemma 6.2.4 trivially hold for that surface. Also, the surface is N-locally-Jordan for all $N \le 5$, and the digital space is N-simply connected for all $N \ge 6$ (Exercise 6.3). However, since there is no N for which both the surface is N-locally-Jordan and the digital space is N-simply connected, we cannot use Lemma 6.2.4 to show that the surface is near-Jordan (which is just as well, since it is not).

Up to now this section has dealt with surfaces in digital spaces in general. Now we turn our attention to boundaries in binary pictures. Before proving our main (and rather important) theorems we need the following technical result.

Lemma 6.2.5. *Let (V, π, f) be a binary picture over the digital space (V, π).*

(i) *Let λ be a tight spel-adjacency in (V, π). Let O be a union of π-components of 1-spels and Q be a λ-component of 0-spels in (V, π, f) such that $S = \partial(O, Q)$ is nonempty. For any c and d in Q, there exists a π-path P from c to d such that $p_S P = 0$.*

(ii) *Let κ be a tight spel-adjacency in (V, π). Let O be a κ-component of 1-spels and Q be a union of π-components of 0-spels in (V, π, f) such that $S = \partial(O, Q)$ is nonempty. For any c and d in O, there exists a π-path P from c to d such that $p_S P = 0$.*

Proof. We prove only (i), since the proof of (ii) is obviously similar.

First, we show that the result is true if $(c, d) \in \lambda$. Since λ is tight, there exists a π-path $\langle c^{(0)}, \cdots, c^{(K)} \rangle$ from c to d such that, for $0 \le k \le K$, either $(c^{(k)}, c) \in \lambda$ or $(c^{(k)}, d) \in \lambda$. Now we prove by induction that, for $0 \le k \le K$,

$$p_S \langle c^{(0)}, \cdots, c^{(k)} \rangle = \begin{cases} 0, & \text{if } c^{(k)} \notin O, \\ 1, & \text{if } c^{(k)} \in O. \end{cases} \qquad (6.2.9)$$

Since $c^{(K)}$ is a 0-spel, and so cannot be in O, this inductive proof meets the aim of this paragraph. Clearly, (6.2.9) is true for $k = 0$, since $c^{(0)}$ is a 0-spel. Now suppose that it is true for some $k - 1$ ($1 \le k \le K$). In the proof we repeatedly use (6.2.4). First we consider the case $c^{(k-1)} \in O$. In this case the first term on the right-hand side of (6.2.4) has value 1, by the induction hypothesis. If $c^{(k)}$ is a 1-spel, then it must also be in O and so the second term on the right-hand side of (6.2.4) has value 0. If $c^{(k)}$ is a 0-spel, then it must be in Q (since either $(c^{(k)}, c) \in \lambda$ or $(c^{(k)}, d) \in \lambda$, and Q is a λ-component of 0-spels that contains both c and d) and so the second term on the right-hand side of (6.2.4) has the value 1. Either way, (6.2.9) is true for k. On the other hand, if $c^{(k-1)} \notin O$, the first term on the right-hand side of (6.2.4) has value 0 by the induction hypothesis. If $c^{(k-1)}$ is a 1-spel, then it cannot be in Q and so the second term on the right-hand side of (6.2.4) has value 0. At the same time $c^{(k)}$ cannot be in O, so (6.2.9) is true for k. If $c^{(k-1)}$ is a 0-spel, then (as before) it must be in Q and so the second term on the right-hand side of (6.2.4) has value 0 if $c^{(k)} \notin O$ and value 1 if $c^{(k)} \in O$. Thus, again (6.2.9) is true in either case for k. This completes the induction proof.

Now consider arbitrary c and d in Q. There exists a λ-path $\langle c^{(0)}, \cdots, c^{(K)} \rangle$ of 0-spels in Q from c to d. By the result in the last paragraph, for $1 \le k \le K$, there exists a π-path $P_k = \langle c^{(k-1)} = c_k^{(0)}, \cdots, c_k^{(m_k)} = c^{(k)} \rangle$ such that $p_S P_k = 0$. Therefore, $P = \langle c_1^{(0)}, \cdots c_1^{(m_1)}, c_2^{(1)}, \cdots, c_2^{(m_2)}, \cdots, c_K^{(1)}, \cdots, c_K^{(m_K)} \rangle$ is a π-path from c to d such that

$$p_S P = \sum_{k=1}^{K} p_S \langle c_k^{(0)}, \cdots, c_k^{(m_k)} \rangle = \sum_{k=1}^{K} p_S P_k = 0, \qquad (6.2.10)$$

as can be shown by repeated application of (6.2.2). (Here $\sum$ refers to modulo 2 additions.) $\square$

Theorem 6.2.6. *Let κ and λ be spel-adjacencies in an N-simply connected digital space (V, π). If at least one of κ and λ is tight, then every N-locally-Jordan $\kappa\lambda$-boundary in a binary picture over (V, π) is near-Jordan.*

Proof. Let O be a κ-component of 1-spels and Q be a λ-component of 0-spels in a binary picture over (V, π) such that $S = \partial(O, Q)$ is an N-locally-Jordan $\kappa\lambda$-boundary. If κ is tight, then (noting that Q has to be a union of π-components of 0-spels) we see from Lemma 6.2.5(ii) that for any c and d in O, there exists a π-path P from c to d such that $p_S P = 0$. Noting that $II(S) \subseteq O$, we see that this implies that condition (i) in Lemma 6.2.4 holds and so S is near-Jordan. The same conclusion can be drawn by a similar argument if λ is tight. $\square$

Theorem 6.2.7. *Let κ and λ be tight spel-adjacencies in an N-simply connected digital space (V, π). A $\kappa\lambda$-boundary in a binary picture over (V, π) is $\kappa\lambda$-Jordan if, and only if, it is N-locally-Jordan.*

Proof. The "only if" part was pointed out just after the definition of N-local-Jordanness. The "if" part follows from Theorem 6.2.6 above and the fact that if κ and λ are tight, then a near-Jordan $\kappa\lambda$-boundary is $\kappa\lambda$-Jordan (Theorem 5.3.8). $\square$

We complete this section by stating an immediate consequence of the definition of a (strong) Jordan pair and the previous theorem. It is hard to overemphasize the importance of this corollary: in the next section it will be used to prove that certain pairs of spel-adjacencies (in particular, pairs that we have used in earlier chapters) are Jordan pairs. (From this will follow proofs of some of our claims in the first two chapters.)

Corollary 6.2.8. *A pair $\{\kappa, \lambda\}$ of tight spel-adjacencies in an N-simply connected digital space (V, π) is a strong Jordan pair (respectively, a Jordan pair) if, and only if, every (respectively, every finite) $\kappa\lambda$-boundary in every binary picture over (V, π) is N-locally-Jordan.*

6.3. Applications to Finding Jordan Pairs

In Chapters 1 and 2 we have claimed that certain boundaries have properties very similar to the properties of simple closed curves in the plane. Our theoretical developments have led us to practically being able to prove the validity of these claims. We need just a few more technical lemmas.

Lemma 6.3.1. *If S is a $\kappa\lambda$-boundary in a binary picture over a digital space (V, π) and $P = \langle a, b, c \rangle$ is a π-path such that $(a, c) \in S$, then $p_S P = 1$.*

Proof. Suppose that $S = \partial(O, Q)$, where O is a κ-component of 1-spels and Q is a λ-component of 0-spels. If b is a 1-spel, then it is in O and $(b, c) \in S$. If b is a 0-spel, then it is in Q and $(a, b) \in S$. In either case, $p_S P = 1$. $\square$

Lemma 6.3.2. *A $\kappa\lambda$-boundary S in a binary picture over a digital space (V, π) is 2-locally-Jordan if $p_S P = 1$ for every π-path $P = \langle c^{(0)}, \cdots, c^{(3)} \rangle$ of length 3 such that $(c^{(0)}, c^{(3)}) \in S$, $c^{(0)} \neq c^{(2)}$, $c^{(1)} \neq c^{(3)}$, $c^{(1)}$ is a 0-spel and $c^{(2)}$ is a 1-spel.*

Proof. Let O be the κ-component of 1-spels and Q be the λ-component of 0-spels such that $S = \partial(O, Q)$. We need to show that $p_S P = 1$ for any π-path $P = \langle c^{(0)}, \cdots, c^{(K)} \rangle$ such that $(c^{(0)}, c^{(K)}) \in S$ and $K = 2$ or 3.

By Lemma 6.3.1, we need only show the result for $K = 3$. If $c^{(0)} = c^{(2)}$ or $c^{(1)} = c^{(3)}$, then it is easy to see that $p_S P = p_S \langle c^{(0)}, c^{(3)} \rangle = 1$. In the following, we assume that both $c^{(0)} \neq c^{(2)}$ and $c^{(1)} \neq c^{(3)}$. There are four possibilities: (i) both $c^{(1)}$ and $c^{(2)}$ are 1-spels, (ii) both $c^{(1)}$ and $c^{(2)}$ are 0-spels, (iii) $c^{(1)}$ is a 1-spel and $c^{(2)}$ is a 0-spel, and (iv) $c^{(1)}$ is a 0-spel and $c^{(2)}$ is a 1-spel. In cases (i), (ii) and (iii), it is easily seen that $p_S P = 1$. That this is also true in case (iv) is exactly the condition in our lemma. $\square$

We call a loop $\langle c^{(0)}, c^{(1)}, c^{(2)}, c^{(3)}, c^{(0)} \rangle$ of length four a *unit square* if both $c^{(0)} \neq c^{(2)}$ and $c^{(1)} \neq c^{(3)}$. An unordered pair $\{\kappa, \lambda\}$ of spel-adjacencies in a digital space is said to be a *normal pair* if, for any unit square $\langle c^{(0)}, c^{(1)}, c^{(2)}, c^{(3)}, c^{(0)} \rangle$, we have $(c^{(0)}, c^{(2)}) \in \kappa$ or $(c^{(1)}, c^{(3)}) \in \lambda$ or both. It is easily seen (by considering the loop $\langle c^{(1)}, c^{(2)}, c^{(3)}, c^{(0)}, c^{(1)} \rangle$) that being normal is indeed a property that does not depend on the order of κ and λ.

Lemma 6.3.3. *If $\{\kappa, \lambda\}$ is a normal pair of spel-adjacencies in a digital space and S is a $\kappa\lambda$-boundary in a binary picture over the digital space, then S is 2-locally-Jordan.*

Proof. By Lemma 6.3.2, we need only show that $p_S P = 1$ for an arbitrary π-path $P = \langle c^{(0)}, \cdots, c^{(3)} \rangle$ of length 3 such that $(c^{(0)}, c^{(3)}) \in S$, $c^{(0)} \neq c^{(2)}$, $c^{(1)} \neq c^{(3)}$, $c^{(1)}$ is a 0-spel and $c^{(2)}$ is a 1-spel. Since $\langle c^{(0)}, c^{(1)}, c^{(2)}, c^{(3)}, c^{(0)} \rangle$ is a loop of length four and both $c^{(0)} \neq c^{(2)}$ and $c^{(1)} \neq c^{(3)}$, we have either $(c^{(0)}, c^{(2)}) \in \kappa$ or $(c^{(1)}, c^{(3)}) \in \lambda$. If $(c^{(0)}, c^{(2)}) \in \kappa$, then $c^{(2)} \in O$, and we see that $p_S P = 1$, whether or not $c^{(1)} \in Q$. If $(c^{(1)}, c^{(3)}) \in \lambda$, then $c^{(1)} \in Q$, and again $p_S P = 1$. $\square$

Theorem 6.3.4. *A normal pair of tight spel-adjacencies in a 2-simply connected digital space is a strong Jordan pair.*

Proof. Let $\{\kappa, \lambda\}$ be such a pair. By the previous lemma, every $\kappa\lambda$-boundary in a binary picture over the digital space is 2-locally-Jordan. By Corollary 6.2.8, it follows that $\{\kappa, \lambda\}$ is a strong Jordan pair. $\square$

Before going on with specifics, let us emphasize the power of this general theorem. Normality is a rather weak (and also very local) restriction on pairs of tight spel-adjacencies. Yet, according to the last theorem, normality is all we need in a 2-simply connected digital space to insure that every (finite or infinite) $\kappa\lambda$-boundary be $\kappa\lambda$-Jordan. Now we turn to applications to digital spaces that are based on grids in N-dimensional euclidean space ($N \geq 1$). We begin with the most frequently used (Z^N, ω_N).

Theorem 6.3.5. *For any positive integer N, (Z^N, ω_N) is 2-simply connected.*

Proof. We show that every loop in (Z^N, ω_N) is 2-equivalent to a trivial loop. We do this by induction on the length of the loop. We have already noted that, in general, every loop of length 1 or 2 is 2-equivalent to a trivial loop (Exercise 6.1). Suppose that every loop in (Z^N, ω_N) of length less than some $K > 2$ is 2-equivalent to a trivial loop. Consider a loop $\langle c^{(0)}, \cdots, c^{(K)} \rangle$ in (Z^N, ω_N) of length K. Now we show that it is 2-equivalent to a loop of length $K - 1$ or $K - 2$ and thus, by the induction hypothesis, is 2-equivalent to a trivial loop.

Since $c^{(1)}$ is ω_N-adjacent to $c^{(0)}$, we have $c^{(1)} \neq c^{(0)}$. (To illustrate our argument, consider Figure 6.1.3. In that figure $c^{(0)} = a = (0, 0)$ and $c^{(1)} = b = (0, 1)$.) Then there is a unique j such that $c_j^{(1)} \neq c_j^{(0)}$. Without loss of generality, assume that $c_j^{(1)} > c_j^{(0)}$. (In Figure 6.1.3, $j = 2$.) Let $z = \max\limits_{1 \leq k \leq K} \left\{ c_j^{(k)} \right\}$. (In Figure 6.1.3, $z = 3$.) Let l be the largest integer in the range $0 < l < K$ such that $c_j^{(l)} = z$. (In Figure 6.1.3, in the left column $l = 5$ with $c^{(5)} = f = (2,3)$ and in the middle column $l = 4$ with $c^{(4)} = e = (1,3)$.) Clearly, $c_j^{(l+1)} = z - 1$. (In Figure 6.1.3, in the left column $c^{(l+1)} = c^{(6)} = g = (2,2)$ and in the middle column $c^{(l+1)} = c^{(5)} = d = (1,2)$.) Let k be the smallest integer such that, for all i in the range $0 < k \leq i \leq l < K$, we have $c_j^{(i)} = z$. (In Figure 6.1.3, $k = 4$ in both the left and the middle columns.) Clearly, $c_j^{(k-1)} = z - 1$. (In Figure 6.1.3, $c^{(k-1)} = c^{(3)} = d = (1,2)$ in both the left and middle columns.) Now we will use induction on $l - k$.

If $l - k = 0$, then $k = l$, and therefore $c^{(k-1)} = c^{(k+1)}$. (This case is illustrated in Figure 6.1.3 by the middle column, for which $k = l = 4$ and $c^{(k-1)} = c^{(k+1)} = d$.) In this case the loop

$$\langle c^{(0)}, \cdots, c^{(k-2)}, c^{(k-1)}, c^{(k)}, c^{(k+1)}, c^{(k+2)}, \cdots, c^{(K)} \rangle \tag{6.3.1}$$

is elementarily 2-equivalent to the loop

$$\langle c^{(0)}, \cdots, c^{(k-2)}, c^{(k-1)} = c^{(k+1)}, c^{(k+2)}, \cdots, c^{(K)} \rangle \tag{6.3.2}$$

which has length $K - 2$, and we are done. (The loop in (6.3.1) is illustrated by the loop of the middle column of Figure 6.1.3, whereas the loop in (6.3.2) is illustrated by the loop of the right column of Figure 6.1.3.)

Now suppose (induction hypothesis) that whenever $l - k = h$, then $\langle c^{(0)}, \cdots, c^{(K)} \rangle$ is 2-equivalent to a loop of length $K - 1$ or $K - 2$. Now we show that the same conclusion holds if $l - k = h + 1$. Let

$$\langle c^{(0)}, c^{(1)}, \cdots, c^{(k)}, \cdots, c^{(l-1)}, c^{(l)}, c^{(l+1)}, \cdots, c^{(K)} \rangle \tag{6.3.3}$$

be a loop. Define j, z, l, and k for this loop as above, and suppose that $l - k = h + 1$. (This is the case in Figure 6.1.3 for the loop associated with the left column with $l = 5$, $k = 4$, and hence $h = 0$. Note also that for this loop $j = 2$ and $z = 3$.) Let $c'^{(l)}$ be a spel such that, for $1 \leq n \leq N$,

$$c_n'^{(l)} = \begin{array}{ll} z - 1, & \text{if } n = j, \\ c_n^{(l-1)}, & \text{otherwise.} \end{array} \tag{6.3.4}$$

(In the left column of Figure 6.1.3, $c'^{(l)} = c'^{(5)} = (1,2) = d$.) Clearly, $c'^{(l)}$ is proto-adjacent to $c^{(l-1)}$. Also, $c^{(l+1)}$ differs from $c^{(l)}$ in exactly the jth component (which is $z - 1$ for the former) and $c^{(l)}$ differs from $c^{(l-1)}$ in exactly one component which is other than the jth. Thus $c'^{(l)}$ is proto-adjacent to $c^{(l+1)}$. (In the left column of Figure 6.1.3, $c^{(l-1)} = c^{(4)} = e$ and $c^{(l+1)} = c^{(6)} = g$, both of which are proto-adjacent to $c'^{(l)} = c'^{(5)} = d$.) It follows that

$$\langle c^{(0)}, c^{(1)}, \cdots, c^{(k)}, \cdots, c^{(l-1)}, c'^{(l)}, c^{(l+1)}, \cdots, c^{(K)} \rangle \tag{6.3.5}$$

is also a loop and is easily seen to be elementarily 2-equivalent to the loop in (6.3.3). (The loop in (6.3.3) is illustrated by the loop of the left column of Figure 6.1.3, whereas the loop in (6.3.5) is illustrated by the loop of the middle column of Figure 6.1.3.) Furthermore, it follows from (6.3.4) that for the loop in (6.3.5) the condition of the induction hypothesis holds. (Indeed, we have already seen that for the middle column of Figure 6.1.3 $l - k = 0 = h$.) So, by the induction hypothesis, the loop in (6.3.5) is 2-equivalent to a loop of length $K - 1$ or $K - 2$. From this it follows that the loop in (6.3.3) is also 2-equivalent to a loop of length $K - 1$ or $K - 2$. $\square$

From the last two theorems we immediately get the following very useful result.

Corollary 6.3.6. *For any positive integer N, a normal pair of tight spel-adjacencies in (Z^N, ω_N) is a strong Jordan pair.*

Now we demonstrate the power of Corollary 6.3.6 by giving some particular examples in (Z^N, ω_N). Later on we show how these examples lead to further strong Jordan pairs. First we need some technical preliminaries.

Consider a digital space (Z^N, ω_N) with N a positive integer. In such a space we call a loop $\langle c^{(0)}, c^{(1)}, c^{(2)}, c^{(3)}, c^{(0)} \rangle$ of length four a *unit lattice square* if the following conditions are satisfied. There exist integers i, j ($1 \leq i \neq j \leq N$) and u, v ($|u| = |v| = 1$) such that

$$c_i^{(1)} = c_i^{(0)} + u, \quad c_j^{(2)} = c_j^{(1)} + v, \quad c_i^{(3)} = c_i^{(2)} - u, \quad c_j^{(0)} = c_j^{(3)} - v. \tag{6.3.6}$$

We refer to the pairs $(c^{(0)}, c^{(2)})$ and $(c^{(1)}, c^{(3)})$ as the *diagonals* of the unit lattice square. (Note that there are no unit lattice squares in (Z^1, ω_1).)

Lemma 6.3.7. *In the digital space (Z^N, ω_N) with $N \geq 1$, a loop is a unit square if, and only if, it is a unit lattice square.*

Proof. It is clear from (6.3.6) that a unit lattice square is a unit square. Conversely, let the loop $\langle c^{(0)}, c^{(1)}, c^{(2)}, c^{(3)}, c^{(0)} \rangle$ be a unit square in (Z^N, ω_N). Since the successive spels are proto-adjacent, they must differ by ± 1 in exactly one coordinate. It is readily verified that since $c^{(0)} \neq c^{(2)}$ and $c^{(1)} \neq c^{(3)}$, these coordinates cannot be the same for two consecutive pairs, and so must be the same (and with opposite signs of the difference) for the alternating pairs. $\square$

Theorem 6.3.8. *For any positive integer N, $\{\delta_N, \omega_N\}$ is a normal pair and hence a strong Jordan pair of spel-adjacencies in (Z^N, ω_N).*

Proof. Let $\langle c^{(0)}, c^{(1)}, c^{(2)}, c^{(3)}, c^{(0)} \rangle$ be a unit square and hence by Lemma 6.3.7 a unit lattice square. Then it is clear from (6.3.6) that both diagonals are in δ_N and so, in particular, $\{\delta_N, \omega_N\}$ is a normal pair. That it is also a strong Jordan pair follows from Corollary 6.3.6, noting the fact that both δ_N and ω_N are tight. $\square$

This theorem, combined with Theorem 5.4.1, immediately yields the validity of two of our early claims.

Corollary 6.3.9. *Claims 1.5.2 and 1.6.2 are valid.*

Lemma 6.3.10. *If κ, λ, κ', λ' are spel-adjacencies in a digital space such that $\{\kappa, \lambda\}$ is normal, $\kappa \subset \kappa'$ and $\lambda \subset \lambda'$, then $\{\kappa', \lambda'\}$ is normal.*

Proof. This follows immediately from the definition of normality on page 143. $\square$

Theorem 6.3.11. *For any positive integer N, if κ and λ are tight spel-adjacencies in (Z^N, ω_N) such that $\delta_N \subset \kappa$, then $\{\kappa, \lambda\}$ is a strong Jordan pair for (Z^N, ω_N).*

Proof. That $\{\kappa, \lambda\}$ is a normal pair follows immediately from Theorem 6.3.8 and the previous lemma, and, consequently, it is also a strong Jordan pair by Corollary 6.3.6. $\square$

As an example, it follows from this theorem that $\{\alpha_N, \omega_N\}$ is a strong Jordan pair of spel-adjacencies for (Z^N, ω_N). However, there are additional examples of strong Jordan pairs of spel-adjacencies which are not special cases of the theorem above. Now we discuss some of these. The first two examples involve the adjacencies β_s, where s is an N-dimensional sign function. Recall that these adjacencies are associated with the hexagonal grid when $N = 2$ and with the fcc grid when $N = 3$. In fact, our third example below states a very desirable property of the fcc grid.

Theorem 6.3.12. *For $N \geq 2$ and for any N-dimensional sign function s, $\{\beta_s, \beta_s\}$ is a normal pair and hence a strong Jordan pair of spel-adjacencies for (Z^N, ω_N).*

Proof. To prove normality, consider any unit square $\langle c^{(0)}, c^{(1)}, c^{(2)}, c^{(3)}, c^{(0)} \rangle$ in (Z^N, ω_N). By Lemma 6.3.7, there exist integers i, j $(1 \leq i \neq j \leq N)$ and u, v $(|u| = |v| = 1)$ such that (6.3.6) is satisfied. It follows that

$$\left(c_i^{(2)} - c_i^{(0)} \right) \times \left(c_j^{(2)} - c_j^{(0)} \right) = u \times v = -\left(c_i^{(3)} - c_i^{(1)} \right) \times \left(c_j^{(3)} - c_j^{(1)} \right) \qquad (6.3.7)$$

and $c_k^{(2)} = c_k^{(0)}$ and $c_k^{(3)} = c_k^{(1)}$, if k is neither i nor j. Since

$$\left| s_{\{i,j\}} \right| = |u \times v| = 1, \qquad (6.3.8)$$

it follows from (3.4.4) and (6.3.7) that exactly one of the diagonals $(c^{(0)}, c^{(2)})$ and $(c^{(1)}, c^{(3)})$ is in β_s. That $\{\beta_s, \beta_s\}$ is also a strong Jordan pair follows from Corollary 6.3.6, noting the fact (Theorem 5.3.5) that β_s is a tight spel-adjacency in (Z^N, ω_N). $\square$

Theorem 6.3.13. *For $N \geq 2$ and for any N-dimensional sign function s, every pair of tight spel-adjacencies in (Z^N, β_s) is a strong Jordan pair for (Z^N, β_s).*

Proof. By Theorem 6.3.12, $\{\beta_s, \beta_s\}$ is a strong Jordan pair of spel-adjacencies in the digital space (Z^N, ω_N). Since β_s is very tight (Theorem 5.3.5), it follows from Theorem 5.5.3 that $\{\beta_s, \beta_s\}$ is also a strong Jordan pair of spel-adjacencies in the digital space (Z^N, β_s). That in fact every pair of tight spel-adjacencies in this space is a strong Jordan pair now follows from Theorem 5.4.6. $\square$

Theorem 6.3.14. *Every pair of tight spel-adjacencies in* (F, β_1) *is a strong Jordan pair for* (F, β_1).

Proof. We know (Theorem 3.4.3) that (F, β_1) is isomorphic to $(Z^3, \beta_{\bar{s}})$, where $\bar{s}$ is a 3-dimensional sign function. Every tight spel-adjacency in (F, β_1) is the isomorphic image of a tight spel-adjacency in $(Z^3, \beta_{\bar{s}})$ (see Exercises 3.11 and 5.9). Now our claim follows from Theorems 6.3.13 and 5.5.4. $\square$

The previous theorem was stated and proved only for the "standard" fcc grid (F, β_1). We leave it to the reader to show (possibly using a "scaling" isomorphism) that the corresponding result is true for any fcc grid (F_ϕ, β_ϕ) with $\phi > 0$. In view of this and of Theorem 5.4.1, we have now delivered on another one of our early promises.

Corollary 6.3.15. *Claim 2.1.1 is valid.*

Being normal is a sufficient condition (in a 2-simply connected digital space) for a pair of tight spel-adjacencies to be strongly Jordan (Theorem 6.3.4) and hence Jordan. However, there are Jordan pairs of spel-adjacencies which are not strongly Jordan and, consequently, not normal. Now we give examples of such Jordan pairs in the digital spaces (Z^N, ω_N) for $N \geq 3$. For each integer n $(1 \leq n \leq N)$, γ_n is that **local** spel-adjacency in (Z^N, ω_N) which satisfies

$$(c, d) \in \gamma_n \Leftrightarrow \sum_{\substack{i=1 \\ i \neq n}}^{N} |d_i - c_i| \leq 1 . \tag{6.3.9}$$

Clearly, each γ_n is a 2-limited (and hence very tight; see Exercise 5.7) spel-adjacency. For any $N \geq 1$, there are N choices for γ_n and, for each of them, every spel has exactly $6N - 4$ spels γ_n-adjacent to it. If $N = 1$ or 2, then each γ_n is the same as α_N. If $N = 3$, the three γ_ns give rise to three "14-adjacencies," each different from one another and from any of the spel-adjacencies discussed in this book so far. It will be seen in Chapter 8 that there are algorithms for tracking finite $\gamma_n \omega_3$-boundaries in (Z^3, ω_3), which operate faster than the algorithms discussed in Chapter 1 for tracking $\delta_3 \omega_3$-boundaries in (Z^3, ω_3). That such algorithms are not only fast, but are also potentially useful, is a consequence of the following result.

Theorem 6.3.16. *For* $N \geq 1$ *and for any integer* n $(1 \leq n \leq N)$, $\{\gamma_n, \omega_N\}$ *is a Jordan pair of spel-adjacencies in* (Z^N, ω_N).

Proof. Let S be an arbitrary finite $\gamma_n \omega_N$-boundary in a binary picture over (Z^N, ω_N). Since γ_n and ω_N are both tight spel-adjacencies in the 2-simply connected digital space (Z^N, ω_N) (see Theorem 6.3.5), all we have to do, according to Corollary 6.2.8, is to show that S is 2-locally-Jordan. We use Lemma 6.3.2. Let $P = \langle c^{(0)}, \cdots, c^{(3)} \rangle$ be a ω_N-path of length 3 such that $(c^{(0)}, c^{(3)}) \in S$, $c^{(0)} \neq c^{(2)}$, $c^{(1)} \neq c^{(3)}$, $c^{(1)}$ is a 0-spel and $c^{(2)}$ is a 1-spel. Now all we need to show is that $p_S P = 1$.

Let O be the γ_n-component of 1-spels and Q be the ω_N-component of 0-spels such that $S = \partial(O, Q)$. Since $\langle c^{(0)}, \cdots, c^{(3)}, c^{(0)} \rangle$ is a unit square and so (by Lemma 6.3.7) a unit lattice square, there exist integers i, j $(1 \leq i \neq j \leq N)$ and u, v $(|u| = |v| = 1)$ such that (6.3.6) is satisfied. If n is one of i or j, then $c^{(2)}$ is γ_n-adjacent to $c^{(0)}$ and, consequently is in O. Under these circumstances, it is easy to see that, whether or not $c^{(1)}$ is in Q, we have $p_S P = 1$. On the other hand, if n is neither i or j, then we also have that either $c^{(2)} \in O$ or $c^{(1)} \in Q$ (and, consequently, $p_S P = 1$), as we prove now by showing that the alternative leads to a contradiction.

Assuming the alternative, we have a loop $\langle c^{(0)}, \cdots, c^{(3)}, c^{(0)} \rangle$ in (Z^N, ω_N), such that $c_n^{(k)}$ is the same for $0 \leq k \leq 3$, and all of the following conditions hold: (i) $(c^{(0)}, c^{(3)}) \in S$, (ii) there exist integers i, j $(1 \leq i \neq j \leq N)$ and u, v $(|u| = |v| = 1)$ such that (6.3.6) is satisfied, and (iii) $c^{(1)}$ is a 0-spel not in Q and $c^{(2)}$ is a 1-spel not in O. Since S is assumed to be finite, we may assume without loss of generality that $\langle c^{(0)}, \cdots, c^{(3)}, c^{(0)} \rangle$ is such a loop for which the common value z of $c_n^{(0)} = c_n^{(1)} = c_n^{(2)} = c_n^{(3)}$ is as great as possible. Now consider the four spels $c'^{(0)}, \cdots, c'^{(3)}$ such that $c'^{(k)}$ is proto-adjacent to $c^{(k)}$ and $c_n'^{(k)} = c_n^{(k)} + 1$ for $0 \leq k \leq 3$. Clearly, $\langle c'^{(0)}, \cdots, c'^{(3)}, c'^{(0)} \rangle$ is a loop in (Z^N, ω_N) such that $c_n'^{(k)}$ is the same (namely, $z+1$) for $0 \leq k \leq 3$. Condition (ii) also holds for this new loop. Looking at condition (iii), we see that $c'^{(1)}$ must be a 0-spel (otherwise $c^{(2)}$ would be in O) and not in Q (otherwise $c^{(1)}$ would be in Q). Similarly, $c'^{(3)}$ must be a 0-spel and hence must be in Q. In view of this, $c'^{(2)}$ must be a 1-spel (otherwise $c'^{(1)}$ would be in Q after all) and not in O (otherwise $c^{(2)}$ would be in O). Similarly, $c'^{(0)}$ must be a 1-spel and hence must be in O. Putting all this together, we see that the loop $\langle c'^{(0)}, \cdots, c'^{(3)}, c'^{(0)} \rangle$ also satisfies conditions (i) and (iii). This contradicts the maximality of z and thus completes the proof. $\square$

The question of whether one could strengthen this theorem by showing that the pairs in question are in fact strong Jordan pairs is answered in the negative by the following counterexample for $\{\gamma_3, \omega_3\}$. Consider the binary picture over (Z^3, ω_3) in which c is a 1-spel if, and only if, $|c_1| + |c_2| = 1$. In this binary picture the infinite set Q of all spels for which $c_1 = c_2 = 0$ is an ω_3-component of 0-spels and the infinite set O of all spels for which $c_1 = 0$ and $c_2 = 1$ is a γ_3-component of 1-spels. The infinite $\gamma_3\omega_3$-boundary between them consists of all surfels of the form $\{(0,1,z), (0,0,z)\}$ and is clearly not near-Jordan. Hence $\{\gamma_3, \omega_3\}$ is not a strong Jordan pair, and hence, by Corollary 6.3.6, it is also not a normal pair of spel-adjacencies in (Z^3, ω_3). (The last part of this statement can also be easily verified by simply using the definition of a normal pair of spel-adjacencies.)

6.4.   1-Simply Connected Digital Spaces

The special class of 1-simply connected digital spaces turns out to be particularly important for a number of reasons. The first one of these is stated in the next theorem: essentially it says that in a 1-simply connected digital space boundaries are automatically Jordan surfaces. In the rest of this section we show that many of the already studied digital spaces are in fact 1-simply connected. Other reasons for paying special attention to this particular family of digital spaces will be seen in the chapters which follow.

Theorem 6.4.1. *Every pair of tight spel-adjacencies in a 1-simply connected digital space* (V, π) *is a strong Jordan pair for* (V, π).

Proof. As a special case of Corollary 6.2.8, a pair $\{\kappa, \lambda\}$ of tight spel-adjacencies in a 1-simply connected digital space (V, π) is a strong Jordan pair if every $\kappa\lambda$-boundary S in

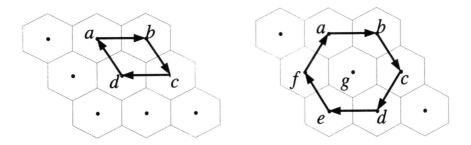

Figure 6.4.1. The hexagonal grid with loops (both equivalent to a trivial loop) indicated.

every binary picture over (V, π) is 1-locally Jordan, i.e., $p_S \langle c^{(0)}, c^{(1)}, c^{(2)} \rangle = 1$, whenever $(c^{(0)}, c^{(2)}) \in S$. However, this is indeed so according to Lemma 6.3.1 $\square$

The digital space represented by Figure 4.1.1 is 1-simply connected. The 1-equivalence of the loop $\langle a, b, c, d, a \rangle$ to a trivial loop can be shown by the following sequence of loops, each one of which is elementarily 1-equivalent to its neighbors in the sequence: $\langle a, b, c, d, a \rangle$, $\langle a, e, b, c, d, a \rangle$, $\langle a, e, c, d, a \rangle$, $\langle a, e, d, a \rangle$, $\langle a \rangle$.

The hexagonal grid with edge-adjacency (see Figure 2.2.3) also gives rise to a 1-simply connected digital space (follows from Theorem 6.4.5 and Exercise 6.6 below). Two examples of loops in this space are shown in Figure 6.4.1. They are both 1-equivalent to a trivial loop, as shown by the sequences $\langle a, b, c, d, a \rangle$, $\langle a, b, d, a \rangle$, $\langle a \rangle$ and $\langle a, b, c, d, e, f, a \rangle$, $\langle a, g, b, c, d, e, f, a \rangle$, $\langle a, g, c, d, e, f, a \rangle$, $\langle a, g, d, e, f, a \rangle$, $\langle a, g, e, f, a \rangle$, $\langle a, g, f, a \rangle$, $\langle a \rangle$.

The square grid with edge-adjacency does not give rise to a 1-simply connected digital space. For otherwise, by Theorem 6.4.1, $\{\omega_2, \omega_2\}$ would be a strong Jordan pair for (Z^2, ω_2), which is not so (Exercise 5.16). The loop $\langle a, b, c, d, a \rangle$, shown in Figure 6.4.2, is not 1-equivalent to a trivial loop. It is 2-equivalent to a trivial loop since the space (Z^2, ω_2) is 2-simply connected (Theorem 6.3.5).

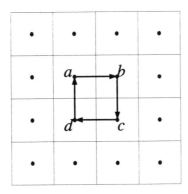

Figure 6.4.2. The square grid gives rise to a 2-simply connected but not 1-simply connected digital space.

Now we come to the basic definition of this section. A spel-adjacency ρ in the digital space (V, π) is said to be *decomposable* if, for all $(c, d) \in \rho$, $\langle c, d \rangle$ is 1-equivalent in (V, ρ) to a π-path. For example, α_N is a decomposable spel-adjacency in $\left(Z^N, \omega_N \right)$.

Lemma 6.4.2.   *If a spel-adjacency ρ in a digital space (V, π) is very tight, then it is decomposable.*

Proof. Suppose that $(c, d) \in \rho$. Since ρ is very tight, there exists a π-path $\langle c^{(0)}, \cdots, c^{(K)} \rangle$ from c to d such that $(c, c^{(k)}) \in \rho$, for $1 \leq k \leq K$. If $K = 0$, then it must be that $c = d = c^{(0)}$ and so $\langle c, d \rangle$ is 1-equivalent in (V, ρ) to the π-path $\langle c^{(0)} \rangle$. Otherwise, define (for $0 \leq k < K$) $P_k = \langle c, c^{(K-k)}, \cdots, c^{(K)} \rangle$. The proof is now completed by noting that each P_k is a ρ-path, $P_0 = \langle c, d \rangle$, P_{K-1} is a π-path and, for $1 \leq k < K$, P_k is elementarily 1-equivalent in (V, ρ) to P_{k-1}. $\square$

In the previous section we have defined what it means that a pair of spel-adjacencies is normal. We will say that ρ is a *normal spel-adjacency* in a digital space (V, π), if $\{\rho, \rho\}$ is a normal pair of spel-adjacencies in (V, π).

Theorem 6.4.3.   *If ρ is a decomposable normal spel-adjacency in a 2-simply connected digital space (V, π), then the digital space (V, ρ) is 1-simply connected.*

Proof. What we need to show is that every loop $P_0 = \langle c^{(0)}, \cdots, c^{(K)} = c^{(0)} \rangle$ in (V, ρ) is 1-equivalent in (V, ρ) to a trivial loop. First we construct a P_K which is a loop in (V, π), and hence in (V, ρ), and is 1-equivalent in (V, ρ) to P_0. We do this in a sequence of K steps. Let $1 \leq k \leq K$, and suppose we already have P_{k-1}. From the fact that ρ is decomposable, we know that there is a π-path that is 1-equivalent in (V, ρ) to $\langle c^{(k-1)}, c^{(k)} \rangle$, and we obtain P_k from P_{k-1} by the replacement of the subsequence $c^{(k-1)}, c^{(k)}$ by the subsequence provided by this π-path. Clearly, P_k is 1-equivalent in (V, ρ) to P_{k-1}. Hence P_K, which is also a loop in (V, π), is 1-equivalent in (V, ρ) to P_0.

Since (V, π) is a 2-simply connected digital space, P_K is 2-equivalent in (V, π) to a trivial loop. All we need to show to complete the proof is that whenever two π-paths are elementarily 2-equivalent in (V, π), then they are 1-equivalent in (V, ρ).

So assume that we have two π-paths as shown in (6.1.1) and (6.1.2) such that $f^{(0)} = d^{(0)}$, $f^{(k)} = d^{(n)}$ and $1 \leq k + n \leq 4$. If in fact $1 \leq k + n \leq 3$, then the two π-paths are elementarily 1-equivalent in (V, π) and hence in (V, ρ), and we are done. For the rest of the proof we assume that $k + n = 4$.

Consider

$$\langle d^{(0)}, \cdots, d^{(n)} = f^{(k)}, \cdots, f^{(0)} = d^{(0)} \rangle . \tag{6.4.1}$$

This is a loop of length four. Its possible manifestations are

(i) $\langle d^{(0)}, d^{(1)}, d^{(2)}, d^{(3)}, d^{(4)} = f^{(0)} = d^{(0)} \rangle$,
(ii) $\langle d^{(0)}, d^{(1)}, d^{(2)}, d^{(3)} = f^{(1)}, f^{(0)} = d^{(0)} \rangle$,
(iii) $\langle d^{(0)}, d^{(1)}, d^{(2)} = f^{(2)}, f^{(1)}, f^{(0)} = d^{(0)} \rangle$,
(iv) $\langle d^{(0)}, d^{(1)} = f^{(3)}, f^{(2)}, f^{(1)}, f^{(0)} = d^{(0)} \rangle$,
(v) $\langle d^{(0)} = f^{(4)}, f^{(3)}, f^{(2)}, f^{(1)}, f^{(0)} = d^{(0)} \rangle$.

It suffices to show that in each of these five cases $\langle d^{(0)}, \cdots, d^{(n)} \rangle$ is 1-equivalent in (V, ρ) to $\langle f^{(0)}, \cdots, f^{(k)} \rangle$ (see the claims made in the paragraph preceding Theorem 6.2.1).

First consider the case when the loop (6.4.1) is not a unit square in (V, π). That means that either the zeroth element in the loop is the same as the second or that the first element in the loop is the same as the third. If both of these elements are ds or both are fs, then our result follows immediately. (For example, if in case (i) $d^{(2)} = d^{(0)}$, then the following sequence of elementary 1-equivalences in (V, ρ) will suffice: $\langle d^{(0)}, d^{(1)}, d^{(2)} = d^{(0)}, d^{(3)}, d^{(4)} = d^{(0)} \rangle$, $\langle d^{(0)}, d^{(3)}, d^{(4)} = d^{(0)} \rangle$, $\langle d^{(0)} = f^{(0)} \rangle$.) The alternative (that one of the two elements is a d and not an f and the other is an f and not a d) can happen only in case (iii), when $d^{(1)} = f^{(1)}$. This, however, implies that $\langle d^{(0)}, d^{(1)}, d^{(2)} \rangle = \langle f^{(0)}, f^{(1)}, f^{(2)} \rangle$, establishing the desired 1-equivalence.

Now assume that the loop (6.4.1) is a unit square in (V, π). The normality of ρ in the digital space (V, π) implies that either the zeroth element in the loop is adjacent in (V, ρ) to the second or that the first element in the loop is adjacent in (V, ρ) to the third. If both of these elements are ds or both are fs, then our result follows immediately. (For example, if in case (i) $(d^{(0)}, d^{(2)}) \in \rho$, then the following sequence of elementary 1-equivalences in (V, ρ) will suffice: $\langle d^{(0)}, d^{(1)}, d^{(2)}, d^{(3)}, d^{(4)} = d^{(0)} \rangle$, $\langle d^{(0)}, d^{(2)}, d^{(3)}, d^{(4)} = d^{(0)} \rangle$, $\langle d^{(0)} = f^{(0)} \rangle$.) The alternative (that one of the two elements is a d and not an f and the other is an f and not a d) can happen only in case (iii), when $(d^{(1)}, f^{(1)}) \in \rho$. In this case the required result follows from the following sequence of elementary 1-equivalences in (V, ρ): $\langle d^{(0)}, d^{(1)}, d^{(2)} \rangle$, $\langle d^{(0)}, d^{(1)}, f^{(1)}, f^{(2)} \rangle$, $\langle f^{(0)}, f^{(1)}, f^{(2)} \rangle$. $\square$

Clearly, this theorem indicates a potentially powerful way of producing examples of 1-simply connected digital spaces. In the next theorem we present several families of 1-simply connected digital spaces, each of which contains N-dimensional examples for arbitrary large N. Due to the preservation of 1-simply connectedness of digital spaces under isomorphism (Exercise 6.6), it is a consequence of the following theorem that the digital spaces obtained by the hexagonal tessellation of the plane with edge-adjacency, by the fcc or bcc grids and face-adjacency of the Voronoi neighborhoods, or by the Khalimsky adjacency of Figure 2.2.5 are all 1-simply connected. First we need a technical lemma.

Lemma 6.4.4. *For every positive integer N, the following spel-adjacencies are normal in $\left(Z^N, \omega_N \right)$: α_N, δ_N, κ_N, ε_e (for any N-dimensional direction vector e) and, when $N \geq 2$, β_s (for any N-dimensional sign function s).*

Proof. We leave it to the reader to show that α_N and δ_N are normal spel-adjacencies in $\left(Z^N, \omega_N \right)$.

To prove the normality of κ_N, let $\langle c^{(0)}, c^{(1)}, c^{(2)}, c^{(3)}, c^{(0)} \rangle$ be a unit square and, hence by Lemma 6.3.7, a unit lattice square. Let i and j be as they occur in (6.3.6). We see that $c^{(0)}$ differs from $c^{(2)}$ only for these two coordinates and that $c^{(1)}$ differs from $c^{(3)}$ also only for these two coordinates. If the parity of $c_i^{(0)}$ is the same as that of $c_j^{(0)}$, then $(c^{(0)}, c^{(2)}) \in \kappa_N$. In the alternative case, since $c^{(1)}$ differs from $c^{(0)}$ only for the ith coordinate and $|c_i^{(1)} - c_i^{(0)}| = 1$, we have that the parity of $c_i^{(1)}$ is the same as that of $c_j^{(1)}$, and so $(c^{(1)}, c^{(3)}) \in \kappa_N$.

To prove the normality of ε_e, consider any unit square $\langle c^{(0)}, c^{(1)}, c^{(2)}, c^{(3)}, c^{(0)} \rangle$. By Lemma 6.3.7, this is a unit lattice square and so satisfies (6.3.6) for some integers i, j $(1 \leq i \neq j \leq N)$ and u, v $(|u| = |v| = 1)$. Since $u = \pm e_i$ and $v = \pm e_j$, it is easy to check that in each of the four possible cases either $(c^{(0)}, c^{(2)}) \in \varepsilon_e$ or $(c^{(1)}, c^{(3)}) \in \varepsilon_e$.

(For example, if $u = -e_i$ and $v = e_j$, then $c_i^{(1)} - c_i^{(3)} = -e_i$, $c_j^{(1)} - c_j^{(3)} = -e_j$, and $c_n^{(1)} - c_n^{(3)} = 0$ for all n other than i or j. This implies that $(c^{(1)}, c^{(3)}) \in \varepsilon_e$, according to the definition of ε_e.)

Finally, the normality of β_s has been proved in Theorem 6.3.12. $\square$

Theorem 6.4.5. *For every positive integer N, the digital space (Z^N, ρ) is 1–simply connected, provided that the proto-adjacency ρ is one of α_N, δ_N, κ_N, ε_e (for any N-dimensional direction vector e) or, when $N \geq 2$, β_s (for any N-dimensional sign function s).*

Proof. According to Theorem 5.3.5, all possible choices of ρ in the statement of the theorem are very tight and hence, according to Lemma 6.4.2, are decomposable spel-adjacencies in (Z^N, ω_N). They are also normal in view of the previous lemma. Therefore the result follows from Theorems 6.3.5 and 6.4.3. $\square$

6.5. Exercises

6.1. Let (V, π) be a digital space and N be a positive integer. Show that every loop of length 1, 2, or 3 in (V, π) is N-equivalent to a trivial loop.

6.2. Let (V, π) be a digital space and N be a positive integer. Prove the following claims.
   (i) If a loop $\langle c^{(0)}, \cdots, c^{(K)} \rangle$ is N-equivalent to a trivial loop, then that trivial loop is $\langle c^{(0)} = c^{(K)} \rangle$.
   (ii) For any π-path $\langle c^{(0)}, \cdots, c^{(K)} \rangle$, the loop $\langle c^{(0)}, \cdots, c^{(K)}, c^{(K)}, \cdots, c^{(0)} \rangle$ is N-equivalent to a trivial loop.
   (iii) For any π-path $\langle c^{(0)}, \cdots, c^{(k)}, \cdots, c^{(K)} \rangle$ and for any loop $\langle d^{(0)}, \cdots, d^{(L)} \rangle$ which is N-equivalent to the trivial loop $\langle c^{(k)} \rangle$, $\langle c^{(0)}, \cdots, c^{(k)}, \cdots, c^{(K)} \rangle$ is N-equivalent to $\langle c^{(0)}, \cdots, c^{(k-1)}, d^{(0)}, \cdots, d^{(L)}, c^{(k+1)}, \cdots, c^{(K)} \rangle$.

6.3. Let (V, π) be a digital space in which V has M elements. Prove that, for any positive integer $N \geq M - 2$, (V, π) is N-simply connected.

6.4. Consider the triangular tessellation of the plane R^2 with edge-adjacency as the proto-adjacency (see Figure 6.5.1). Show that this digital space is 4-simply connected.

6.5. Prove that, for $1 \leq N \leq 3$ and for any local spel-adjacency ρ in the digital space (Z^N, ω_N), (Z^N, ρ) is a 2-simply connected digital space.

6.6. Let N be a positive integer. Show that if two digital spaces are isomorphic, then one of them is N-simply connected if, and only if, the other one is N-simply connected.

6.7. Let S be an antisymmetric surface and (c, d) be a surfel in a digital space (V, π). Show that (c, d) crosses S if, and only if, the π-path $\langle c, d \rangle$ crosses S.

6.8. Verify the four claims made in the paragraph preceding Theorem 6.2.1.

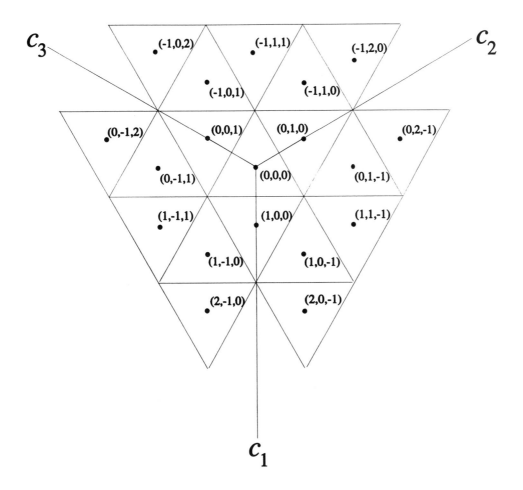

Figure 6.5.1. The triangular tessellation of R^2. Triangles are considered adjacent if they share an edge. The triangles can be represented by elements of Z^3 by using coordinate axes as indicated. Only those elements c of Z^3 for which the sum of the components is either 0 or 1 represent a triangle. Triangles represented by vectors $c, d \in Z^3$ are adjacent if, and only if, $|c_1 - d_1| + |c_2 - d_2| + |c_3 - d_3| = 1$.

6.9. Show that, for any positive integer N, $\{\alpha_N, \alpha_N\}$ is a strong Jordan pair of spel-adjacencies in (Z^N, ω_N).

6.10. Show that, for any $\phi > 0$, every pair of tight spel-adjacencies in (F_ϕ, β_ϕ) is a strong Jordan pair for (F_ϕ, β_ϕ).

6.11. Show that, for any $N \geq 1$, there are N choices for γ_n and, for each of them, every spel has exactly $6N - 4$ spels γ_n-adjacent to it.

6.12. According to the previous exercise, there are fourteen spels γ_1-adjacent to a given spel in Z^3. We also know (see Chapter 2) that there are fourteen voxels face-adjacent to a given voxel in the bcc grid. Show that, nevertheless, (Z^3, γ_1) is not isomorphic to the bcc grid with face-adjacency.

6.13. Show for the ϕ defined on page 42 that $\{\phi, \omega_4\}$ is a Jordan pair of spel-adjacencies in (Z^4, ω_4).

6.14. Show that not every tight spel-adjacency is decomposable.

6.15. Show that α_N and δ_N are normal spel-adjacencies in (Z^N, ω_N).

6.16. Let a loop $\langle c^{(0)}, c^{(1)}, \cdots, c^{(K-1)}, c^{(K)} = c^{(0)} \rangle$ in a digital space (V, π) be called a *cycle* if its length K is at least 3 and $c^{(i)} \neq c^{(j)}$, for $0 \leq i < j < K$. A π-path $\langle d^{(0)}, \cdots, d^{(L)} \rangle$ is said to be a *bridge* of this cycle, if for some i and j, $0 \leq i < j < K$, $d^{(0)} = c^{(i)}$, $d^{(L)} = c^{(j)}$, and both $\langle c^{(0)}, \cdots, c^{(i-1)}, c^{(i)} = d^{(0)}, d^{(1)}, \cdots, d^{(L-1)}, d^{(L)} = c^{(j)}, c^{(j+1)}, \cdots, c^{(K)} = c^{(0)} \rangle$ and $\langle c^{(j)} = d^{(L)}, d^{(L-1)}, \cdots, d^{(1)}, d^{(0)} = c^{(i)}, c^{(i+1)}, \cdots, c^{(j-1)}, c^{(j)} \rangle$ are cycles in (V, π) of length less than K. (V, π) is said to be a *bridged digital space* (see, e.g., [9]) if every cycle of length at least 4 has a bridge.

   (i) Show that the digital space represented in Figure 4.1.1 is not bridged.

   (ii) Demonstrate that the two cycles in the digital space (H, ε) which are shown in Figure 6.4.1 both have bridges.

6.17. Using the definitions of the previous exercise, show that every bridged digital space is 1-simply connected. (*Hint:* you may wish to use the facts stated in Exercise 6.2.)

7
Jordan Graphs

"And stretch de boundary line to de oder side ob Jordan."

Jordan is a Hard Road to Travel, a song by D. D. Emmett.

7.1. The Theory of (Strong) Jordan Graphs

In the previous chapter we have pointed out the existence of digital spaces (V, π) for which $\{\pi, \pi\}$ is a strong Jordan pair. According to Theorem 6.4.1, all 1-simply connected digital spaces are in this category, including by Theorem 6.4.5 (Z^N, α_N), (Z^N, δ_N), (Z^N, κ_N), (Z^N, ε_e) (for any N-dimensional direction vector e) and, when $N \geq 2$, (Z^N, β_s), for any N-dimensional sign function s. That the digital spaces of this last kind have the stated property also follows directly from Theorem 6.3.13. We have also shown (in Theorem 6.3.14) that the fcc grid with face-adjacency (i.e., (F, β_1)) also has the stated property. Following [2], we refer to digital spaces with this property as strong Jordan graphs.

A digital space (V, π) is called a *(strong) Jordan graph* if $\{\pi, \pi\}$ is a (strong) Jordan pair for (V, π). We do not bother to define and to discuss the notion of a weak Jordan graph, since in view of Theorem 5.4.8 such a notion would coincide with that of a Jordan graph in all practically interesting situations. The following is an immediate consequence of Theorem 5.4.6.

Corollary 7.1.1. *In a (strong) Jordan graph every pair of tight spel-adjacencies is a (strong) Jordan pair. In particular, if ρ is a tight spel-adjacency in a (strong) Jordan graph (V, π), then (V, ρ) is also a (strong) Jordan graph.*

This corollary, combined with Theorem 5.4.1, indicates why (strong) Jordan graphs are so important: if (V, π) is a Jordan graph and κ and λ are tight spel-adjacencies in it, then every finite $\kappa\lambda$-boundary in every binary picture over (V, π) has the Jordan properties listed in Theorem 5.4.1. If (V, π) is a strong Jordan graph, then we do not even need the finiteness of the $\kappa\lambda$-boundary. In particular, these statements are valid in the special case of $\kappa = \lambda$, and so for (strong) Jordan graphs we do not have to use different adjacencies for the 1-spels and the 0-spels to ensure that boundaries are Jordan surfaces. Furthermore, we can formally restate the following immediate consequence of Theorem 6.4.1.

Corollary 7.1.2. *Every 1-simply connected digital space is a strong Jordan graph.*

By Theorem 6.4.5, this implies that (Z^N, α_N), (Z^N, δ_N), (Z^N, κ_N), (Z^N, ε_e) (for any N-dimensional direction vector e) and, when $N \geq 2$, (Z^N, β_s) (for any N-dimensional sign function s) are all strong Jordan graphs. Using various isomorphisms discussed in Chapter 3 together with the following immediate corollary of Theorem 5.5.4, it follows that the hexagonal grid with edge-adjacency and either the fcc grid or the bcc grid with face-adjacency also give rise to strong Jordan graphs.

Corollary 7.1.3. *If two digital spaces are isomorphic, then one is a strong Jordan graph (respectively, a Jordan graph) if, and only if, the other is a strong Jordan graph (respectively, a Jordan graph).*

Not all spaces of interest are strong Jordan graphs: in fact, for $N \geq 2$, (Z^N, ω_N) is not even a Jordan graph; see Exercise 5.16. We postpone till Theorem 7.1.9 below the demonstration that not all Jordan graphs are strong Jordan graphs. Before going any further in our development, we discuss why the Corollaries 7.1.1 and 7.1.2 are as strong as they can possibly be.

First we demonstrate that the condition of tightness is necessary in Corollary 7.1.1. Consider the following one-dimensional example. For any positive integer i, let the adjacency ν_i on Z be defined by (5.3.5). It is trivial to prove that (Z, ν_1) is a strong Jordan graph (in fact every $\nu_1\nu_1$-boundary has exactly one element in this digital space) and also that ν_2 and ν_3 are tight spel-adjacencies in (Z, ν_1). It follows from Corollary 7.1.1 that (Z, ν_2) and (Z, ν_3) are also strong Jordan graphs. On the other hand, ν_4 is not a tight spel-adjacency in (Z, ν_1), and (Z, ν_4) is not a Jordan graph. This can be seen from Figure 7.1.1, in which the heavily shaded region is a ν_4-component of 1-spels, the lightly shaded region is a ν_4-component of 0-spels (consisting of one spel), and the $\nu_4\nu_4$-boundary between them is $\{(9, 10), (14, 10)\}$. The ν_4-path $\langle 9, 13, 12, 11, 10 \rangle$ connects the immediate interior of this $\nu_4\nu_4$-boundary to its immediate exterior without crossing it.

Corollary 7.1.2 is also as strong as it can possibly be since in its statement we cannot replace 1-simply connected by 2-simply connected. This is so since (Z^2, ω_2) is 2-simply connected (Theorem 6.3.5) and, as we have already repeatedly pointed out, it is not a Jordan graph. Also, we show below that the converse of Corollary 7.1.2 is not valid: we will give an example of a strong Jordan graph which is not a 1-simply connected digital space. To do this, however, we need to make a further excursion into graph theory. This excursion turns out to be very useful. In addition to yielding the promised example of a strong Jordan

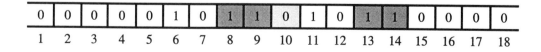

Figure 7.1.1. Illustration that (Z, ν_4) is not a Jordan graph.

graph which is not a 1-simply connected digital space, it will also give us new and interesting characterizations of the notions of strong Jordan pairs and of Jordan pairs.

Our previous excursion into graph theory in Chapter 1 concentrated on digraphs. Now we introduce the notion of a *graph* as an ordered pair (M, ρ), where M is a nonempty set (we refer to its elements as *nodes*) and ρ is an irreflexive symmetric binary relation on M (we refer to unordered pairs $\{c, d\}$ as *arcs*, if $(c, d) \in \rho$). The differences which distinguish this notion of a graph from the notion of a digraph (as we have defined them) are the following:

(i) M need not be finite;

(ii) ρ has to be symmetric;

(iii) arcs are unordered pairs of elements of M.

Otherwise, we use the notation and terminology of the section on digraphs in Chapter 1. A concept that we have not defined for digraphs (although we could have, but it was not necessary) is the following: a *cycle* in a graph is a path $\langle c^{(0)}, \cdots, c^{(K)} \rangle$ such that $K \geq 3$ and, for $0 \leq i < j \leq K$, $c^{(i)} = c^{(j)}$ if, and only if, $i = 0$ and $j = K$. We say that (M, ρ) is an *acyclic graph* if there are no cycles in it.

As an example, we represent two graphs in Figure 7.1.2. For both of them the set of nodes is {O1, O2, O3, I1, I2, I3}, and the set of arcs is indicated by the lines in the figure. Neither of these graphs is acyclic because the cycle $\langle O2, I2, O3, I3, O2 \rangle$ is in both of them. However, if we removed {O2, I3} from the set of arcs, the graph represented on the left would become acyclic, whereas the graph represented on the right would not do so. Both of these graphs are a special type, as we explain now.

Let κ and λ be tight spel-adjacencies in the digital space (V, π). The $\kappa\lambda$-*adjacency graph* of a binary picture (V, π, f) has as its nodes the κ-components of the 1-spels and the λ-components of the 0-spels, with an arc between two such nodes O and Q if, and only if, $\partial(O, Q)$ is nonempty. Consider, for example, Figure 7.1.1. There are three ν_4-components of 0-spels (O1 = {$\cdots$, 1, 2, 3, 4, 5, 7}, O2 = {10}, and O3 = {12, 15, 16, 17, 18, $\cdots$}) and three ν_4-components of 1-spels (I1 = {6}, I2 = {8, 9, 13, 14}, and I3 = {11}). The corresponding $\nu_4\nu_4$-adjacency graphs in (Z, ν_1) and in (Z, ν_4) are shown in Figure 7.1.2(a) and (b), respectively.

Our next theorem characterizes strong Jordan pairs in terms of adjacency graphs. Its proof uses the following technical lemma.

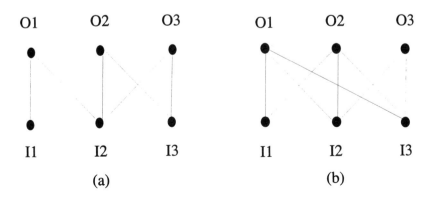

Figure 7.1.2. $\nu_4\nu_4$-adjacency graphs associated with Figure 7.1.1: (a) in (Z, ν_1) and (b) in (Z, ν_4).

Lemma 7.1.4.   *Let* (V, π, f) *be a binary picture.*

(i) *If* κ *is a tight spel-adjacency in* (V, π) *and* O *is a* κ-*component of the set of 1-spels, then for every pair of elements of* O, *there is a* π-*path from one to the other such that all 1-spels occurring in this* π-*path are also in* O.

(ii) *If* λ *is a tight spel-adjacency in* (V, π) *and* Q *is a* λ-*component of the set of 0-spels, then for every pair of elements of* Q, *there is a* π-*path from one to the other such that all 0-spels occurring in this* π-*path are also in* Q.

Proof. We prove only (i), since the proof of (ii) is strictly analogous. Let (V, π, f) be a binary picture, κ be a tight spel-adjacency in (V, π) and O be a κ-component of the set of 1-spels. Since O is κ-connected, we can go from any element in it to any other element in it by a κ-path entirely in O. This κ-path can be used to create the desired π-path, provided only that we can show that, for any κ-adjacent c and d in O, there is a π-path connecting them such that all 1-spels in this π-path are also in O. Since κ is tight, there exists a π-path $\langle c^{(0)}, \cdots, c^{(K)} \rangle$ from c to d such that, for $0 \le k \le K$ either $(c, c^{(k)}) \in \kappa$, or $(d, c^{(k)}) \in \kappa$, or both. Since O is a κ-component of the set of 1-spels, the required result follows. $\square$

For an illustration of Lemma 7.1.4(ii), consider Figure 5.4.4. The lightly shaded δ-component of the 0-spels is not ω_2-connected, and so there are pairs of elements in this δ-component which cannot be connected by a ω_2-path which does not leave the δ-component. However, for any pair of elements of the δ-component, there is a connecting ω_2-path all of whose 0-spels are in the δ-component.

Theorem 7.1.5.   *For tight spel-adjacencies* κ *and* λ *in the digital space* (V, π), $\{\kappa, \lambda\}$ *is a strong Jordan pair for* (V, π) *if, and only if, the* $\kappa\lambda$-*adjacency graph of any binary picture over* (V, π) *is acyclic.*

Proof. First suppose that $\{\kappa, \lambda\}$ is not a strong Jordan pair for (V, π). Then there is a binary picture (V, π, f) with a κ-component O of the 1-spels and a λ-component Q of the 0-spels such that there is a π-path $\langle c^{(0)}, \cdots, c^{(K)} \rangle$ not crossing $\partial(O, Q)$ from an element of $II(\partial(O, Q))$ to an element of $IE(\partial(O, Q))$. Clearly, there is an arc between O and Q in the $\kappa\lambda$-adjacency graph of (V, π, f). This arc can be shown to be an arc in a cycle in this graph by following the π-path $\langle c^{(0)}, \cdots, c^{(K)} \rangle$. The fact that this π-path begins in O and ends in Q but does not cross $\partial(O, Q)$ has the following two consequences: (i) the 0-spel in the π-path that follows the last spel in it which is also in O must be in a λ-component Q' of the 0-spels other than Q and (ii) the 1-spel in the π-path that precedes the first spel in it which is also in Q must be in a κ-component O' of the 1-spels other than O. It follows that there are arcs between O and Q' and between O' and Q in the $\kappa\lambda$-adjacency graph of (V, π, f). We get a cycle $\langle O, Q', \cdots, O', Q, O \rangle$ in the $\kappa\lambda$-adjacency graph of (V, π, f) by putting in place of the "$\cdots$" all the κ-components of the 1-spels and λ-components of the 0-spels visited by the π-path between Q' and O' (eliminating any repeated visits that may occur).

Now suppose that there is a binary picture (V, π, f) such that its $\kappa\lambda$-adjacency graph contains the cycle $\langle O, Q', \cdots, O', Q, O \rangle$, where O is a κ-component of the 1-spels and Q is a λ-component of the set of 0-spels. To show that $\{\kappa, \lambda\}$ is not a strong Jordan pair for (V, π), it is sufficient to show that $\partial(O, Q)$ is not near-Jordan. By Lemma 3.2.2 and Theorem 5.3.6, it is sufficient to show that there is a π-path from a spel in O to a spel in Q which does not cross $\partial(O, Q)$. Let $(c, d') \in \partial(O, Q')$ and $(c', d) \in \partial(O', Q)$. Since O and O' are different κ-components of the 1-spels and Q and Q' are different λ-components of the 0-spels, neither of these surfels is in $\partial(O, Q)$. Now we discuss how to fill in the

"···" in the π-path $\langle c, d', \cdots, c', d \rangle$ from an element of O to an element of Q so that it does not cross $\partial(O, Q)$. Let e' be an element of Q' which is π-adjacent to some element of the κ-component of the set of 1-spels that follows Q' in the cycle whose existence we are assuming. (This κ-component of the set of 1-spels could be, but does not have to be, O'.) By Lemma 7.1.4(ii), there is a π-path from d' to e' such that all 0-spels in this π-path are also in Q'. Since Q and Q' are different λ-components of the 0-spels, this π-path does not cross $\partial(O, Q)$. Similarly, we can extend the π-path to an element of the κ-component of the 1-spels that follows Q' in the cycle without crossing $\partial(O, Q)$. This way we have moved from Q' in the cycle to the node of the $\kappa\lambda$-adjacency graph that follows it in the cycle. By repeating this process, using alternatively Lemma 7.1.4(i) and Lemma 7.1.4(ii) (and noting that until and including the point when we have taken care of O', the nodes with which we are dealing are different from O and from Q), we can generate the required π-path from c to d. $\square$

This theorem is of general interest since it gives a new characterization of the concept of a strong Jordan pair. From the point of view of the current chapter, its relevance lies in the fact that, in particular, we can conclude the following from it.

Corollary 7.1.6. *A digital space* (V, π) *is a strong Jordan graph if, and only if, the* $\pi\pi$-*adjacency graph of any binary picture is acyclic.*

As an illustration consider Figure 4.1.1. The spel e is adjacent to the other four spels. Hence, in any binary picture in which e is a 1-spel, there is at most one π-component of the set of 1-spels, and in any binary picture in which e is a 0-spel, there is at most one π-component of the set of 0-spels. In either case, the $\pi\pi$-adjacency graph is acyclic, and so, by Corollary 7.1.6, the digital space is a strong Jordan graph. (This was a pretty nifty way of showing that this is a strong Jordan graph, as opposed to looking at the 32 possible ways of assigning 0s and 1s to the five spels.)

As a second application now we give the promised example (originally due to [8]) of a strong Jordan graph which is not a 1-simply connected digital space. The set of spels is $V = \{0, 1, 2, 3, 4, 5, 6\}$, and two spels are π-adjacent if, and only if, the modulo 7 difference between them is ± 1 or ± 2. We illustrate (V, π) in Fig 7.1.3, where the arrowed lines indicate the proto-adjacencies.

First we show that this digital space (V, π) is a strong Jordan graph. By Corollary 7.1.6, it is sufficient to show that the $\pi\pi$-adjacency graph of any binary picture over (V, π) is acyclic. We assume the opposite and show that this leads to a contradiction. If there is a cycle in the $\pi\pi$-adjacency graph of a binary picture over (V, π), then there must be at least two distinct π-components of the 1-spels in this binary picture. Without loss of generality (since the space is isomorphic to itself under "rotation"; see Figure 7.1.3), we may assume that 0 is a 1-spel. Then the other component of the set of 1-spels must contain at least one of 3 or 4 (all other spels being proto-adjacent to 0). Assume that this other component of the set of 1-spels contains 3 (were we to assume that it contained 4, the argument would be strictly analogous). In Figure 7.1.3, we have marked these two 1-spels by the heavier shaded squares. The spels marked by the lighter shaded squares (namely, 1, 2 and 5), must be 0-spels, for otherwise 0 and 3 would be in the same component. By the assumption of the existence of a cycle in the $\pi\pi$-adjacency graph, there must also be at least two distinct π-components of the 0-spels in the binary picture. This can only happen if both 4 and 6 are

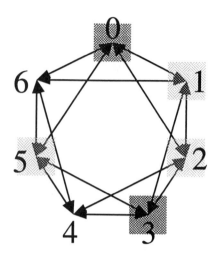

Figure 7.1.3. A strong Jordan graph which is not a 1-simply connected digital space.

1-spels. That indicates that, after all, the binary picture has only one π-component of 1-spels. This is the promised contradiction.

Finally, we show that (V, π) of Figure 7.1.3 is not 1-simply connected. Consider the surface $S = \{(0, 2), (1, 2), (1, 3)\}$. This surface is 1-locally-Jordan. (By the definition of an N-locally-Jordan surface, to prove that S is a 1-locally-Jordan surface, we have to show that, for any π-path $P = \langle c^{(0)}, c^{(1)}, c^{(2)} \rangle$ such that $\left(c^{(0)}, c^{(2)} \right) \in S$, $p_S(P) = 1$. Since S is so small, we can list all such possible π-paths. They are $\langle 0, 1, 2 \rangle$, $\langle 1, 0, 2 \rangle$, $\langle 1, 3, 2 \rangle$, $\langle 1, 2, 3 \rangle$, and the crossing parity of each one of them through S is odd.) Now consider the loop $\langle 0, 2, 4, 6, 0 \rangle$ in (V, π). Its crossing parity through S is odd and so, by Lemma 6.2.3, the same is true for any loop 1-equivalent to it. It follows that $\langle 0, 2, 4, 6, 0 \rangle$ is not 1-equivalent to a trivial loop (whose crossing parity has to be even), and so (V, π) is not 1-simply connected.

Before going further, consider for a moment what we have just done in the previous paragraphs. We had a very specific task to perform: the illustration that there exists a strong Jordan graph which is not a 1-simply connected digital space. Our proof that the digital space represented in Figure 7.1.3 is an example of this has demonstrated what we promised in Chapter 1 as something that we will be able to do, namely, the straightforward validation of a claim by appealing to some general results (in this case Corollary 7.1.6 and Lemma 6.2.3), rather than by providing a long-winded proof cooked up specifically for it. Therefore it appears that we are making progress toward our grand aim of providing a useful theory of digital geometry. So let us go on.

Somewhat oddly, the characterization of Jordan pairs (as opposed to strong Jordan pairs) requires some additional machinery: we need to concentrate our attention on finitary spel-adjacencies of which at least one is very tight. However, the discussion up to now in this book clearly indicates that from the practical point of view such a restriction is of no significance: just about all spel-adjacencies that we may desire to use in practice are finitary and very tight.

Theorem 7.1.7. *If κ and λ are finitary spel-adjacencies in the digital space (V, π) such that at least one of them is very tight, then $\{\kappa, \lambda\}$ is a Jordan pair if, and only if, the $\kappa\lambda$-adjacency graph of any finite picture is acyclic.*

Proof. First suppose that $\{\kappa, \lambda\}$ is not a Jordan pair for (V, π). We assume that κ is very tight; the proof in the alternative case is analogous. There is a finite $\kappa\lambda$-boundary S in a binary picture (V, π, g), such that S is not near-Jordan. By Lemma 5.4.7, S is also a $\kappa\lambda$-boundary in some finite binary picture (V, π, f). Now we can show that the $\kappa\lambda$-adjacency graph of (V, π, f) is cyclic, just as we have done in the first half of the proof of Theorem 7.1.5.

Conversely, suppose that the $\kappa\lambda$-adjacency graph of a finite picture is not acyclic Then we can generate a $\kappa\lambda$-boundary in this finite picture which is not near-Jordan, just as we have done in the second half of the proof of Theorem 7.1.5. In addition we observe that finiteness of the picture together with the finitariness of π (π has to be finitary since it is a subset of both κ and λ) implies that this not near-Jordan $\kappa\lambda$-boundary is finite. This shows that $\{\kappa, \lambda\}$ is not a Jordan pair for (V, π). $\square$

Corollary 7.1.8. *A finitary digital space (V, π) is a Jordan graph if, and only if, the $\pi\pi$-adjacency graph of each finite picture over (V, π) is acyclic.*

Note that neither the statements nor the proofs of Lemma 7.1.4, Theorem 7.1.5 and the previous theorem depend on the concept of a Jordan graph. They could have been presented in Chapter 5, but we have chosen to present them here since our main motivation for proving them is the characterization of (strong) Jordan graphs stated in Corollaries 7.1.6 and 7.1.8. Of particular interest is the fact that the proof of the previous theorem makes essential use of Lemma 5.4.7: this provides additional motivation for considering the concept of a weak Jordan pair.

Now we return to our promised example of a Jordan graph which is not a strong Jordan graph. Let

$$T = \{ c \in Z^2 \,|\, 0 \le c_1 \le 3 \}. \tag{7.1.1}$$

This can be thought of as a four-pixel wide vertical band in Z^2. However, using the hexagonal interpretation of pairs of integers as represented in Figure 2.2.3, we can also think of elements of T as representing the set of hexagons shown in Figure 7.1.4: an infinitely long band in the c_2-direction, but only four pixels wide in the c_1-direction. On this set, we define the adjacency

$$\begin{aligned}
\tau = \{ \,(c, d) \,|\, (c, d) \in \alpha_2 \ \& \ (c_1 - d_1) \neq (d_2 - c_2) \\
\text{or } |c_1 - d_1| = 3 \ \& \ c_2 = d_2 \\
\text{or } |c_1 - d_1| = 3 \ \& \ c_2 - d_2 = (d_1 - c_1)/3 \,\}.
\end{aligned} \tag{7.1.2}$$

The first line of this definition coincides exactly with the definition of β on page 50. The other two lines are relevant only when the two spels are at opposite edges of the band.

It follows that in the digital space (T, τ) the spels proto-adjacent to a spel c for which $1 \le c_1 \le 2$ are exactly the six spels β-adjacent to it. From the isomorphism (discussed in Chapter 3) between (Z^2, β) and (H, ε), we see that these are the six (hexagonal) pixels edge-adjacent to the pixel corresponding to c in Figure 7.1.4. Thus the spels proto-adjacent to $(2, -1)$ are $(2, 0)$, $(1, -1)$, $(1, -2)$, $(2, -2)$, $(3, -1)$, and $(3, 0)$. For the other spels

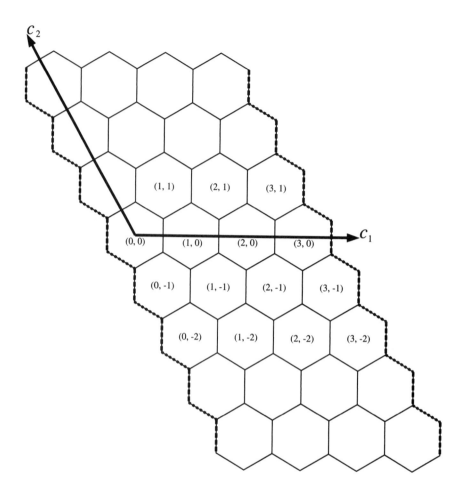

Figure 7.1.4. Representation of the digital space (T, τ).

(which are on the edge of the band), there are also six spels adjacent to them: in addition to the four spels provided by the edge-adjacency between hexagons, we get two more spels which can be thought of as those provided by the hexagons in Figure 7.1.4 if we wrap this space around so that the broken lines are attached to the broken lines and the dotted lines are attached to the dotted lines. Thus the spels proto-adjacent to $(3, -1)$ are $(3, 0)$, $(2, -1)$, $(2, -2)$, $(3, -2)$, $(0, -1)$, and $(0, 0)$. The last two of these are provided by the second and the third lines, respectively, of definition (7.1.2). Since every spel has exactly six spels proto-adjacent to it, τ is a finitary binary relation on T.

Theorem 7.1.9. *Not every Jordan graph is a strong Jordan graph.*

Proof. Consider that binary picture (T, τ, t) over the just defined digital space (T, τ), for which

$$t(c) = \begin{array}{l} 0, \text{ if } c_1 \text{ is even,} \\ 1, \text{ if } c_1 \text{ is odd .} \end{array} \qquad (7.1.3)$$

There are four nodes in the $\tau\tau$-adjacency graph of this binary picture; they are

$$N_i = \{ \, c \mid c_1 = i \, \}, \text{ for } 0 \le i \le 3 \,. \tag{7.1.4}$$

There is an arc between nodes N_i and N_j if, and only if, the modulo 4 difference between i and j is ± 1. Hence $\langle N_0, N_1, N_2, N_3, N_0 \rangle$ is a cycle in the $\tau\tau$-adjacency graph of (T, τ, t) and so, by Corollary 7.1.6, (T, τ) is not a strong Jordan graph.

To show that it is a Jordan graph, we demonstrate that assuming that it is not a Jordan graph implies that $(\mathbb{Z}^2, \beta)$ is also not a Jordan graph, which we know is false (see the paragraph after Corollary 7.1.2). If (T, τ) is not a Jordan graph, then (by Lemma 5.4.7, recalling that τ is finitary) there is a finite picture (T, τ, f) with a τ-component O of the 1-spels and a τ-component Q of the 0-spels such that there is a τ-path not crossing $\partial(O, Q)$ from a spel $c \in II(\partial(O, Q))$ to a spel $d \in IE(\partial(O, Q))$. Now we give a construction which will associate with any such picture a finite picture $(\mathbb{Z}^2, \beta, g)$ such that if O' is the β-component of the set of 1-spels containing c and Q' is the β-component of the set of 0-spels containing

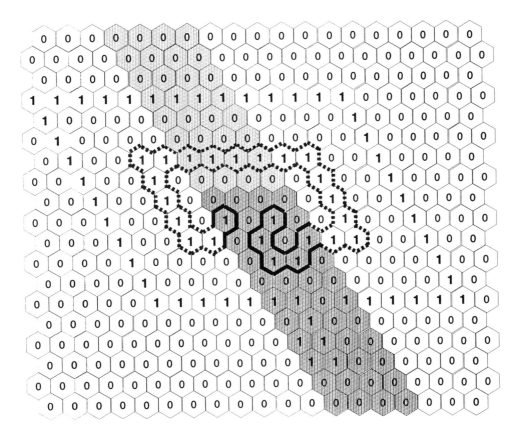

Figure 7.1.5. Illustration of the construction to show that (T, τ) is a Jordan graph. T is the shaded area, and D is the more heavily shaded part of it. The unbroken heavy lines (with orientation from 1-spels to 0-spels) form a $\tau\tau$-boundary in (T, τ) that is extended by the construction into a $\beta\beta$-boundary in $(\mathbb{Z}^2, \beta)$, as indicated by the broken heavy lines (with orientation from 1-spels to 0-spels).

d, then there is a β-path from c to d not crossing the $\beta\beta$-boundary $\partial(O', Q')$. This would indicate that (Z^2, β) is not a Jordan graph.

The construction is done as follows. Since (T, τ, f) is finite, there exists an integer u such that $f(c) = 0$ if $c_2 \geq u$. We define

$$D = \{\, c \mid 0 \leq c_1 \leq 3,\ c_2 \leq u \,\}, \qquad (7.1.5)$$

and we set g equal to f for spels in D (which is a subset of both Z^2 and of T). Outside D we set g equal to 0 with the following exception. For every $c \in D$ such that $c_1 = 3$ and $f(c) = 1$, we assign the value 1 to $g(d)$ if either $4 \leq d_1 \leq 3 + u - c_2$ and $d_2 = c_2$, or $d_1 = 3 + u - c_2$ and $c_2 \leq d_2 \leq 2u - c_2$, or $-u + c_2 \leq d_1 \leq 3 + u - c_2$ and $d_2 = 2u - c_2$, or $d_1 = -u + c_2$ and $c_2 \leq d_2 \leq 2u - c_2$, or $-u + c_2 \leq d_1 \leq -1$ and $d_2 = c_2$. This is done to simulate the wraparound in the digital space (T, τ) by "corridors" (of all 1-spels or all 0-spels) outside D in (Z^2, β). We illustrate the process in Figure 7.1.5. We leave it to the reader to check that the construction has the following properties. If O' is the β-component of the set of 1-spels containing c and Q' is the β-component of the set of 0-spels containing d, then $O \cap D = O' \cap D$ and $Q \cap D = Q' \cap D$. We can simulate the τ-path from c to d not crossing $\partial(O, Q)$ by a β-path not crossing $\partial(O', Q')$, by following its moves inside D and using the "corridors" whenever the τ-path wraps around. (The connected set $\{\, c \mid 0 \leq c_1 \leq 3,\ c_2 = u \,\}$ of 0-spels plays a special role in this simulation.) $\square$

7.2. Jordan Surfaces

In Chapter 3 we introduced the notion of a Jordan surface in a digital space (V, π) as referring to those surfaces which are minimally near-Jordan or (equivalently by Theorem 3.3.5) $\pi\pi$-Jordan. The terminology is justified by Corollary 3.3.6, which states that Jordan surfaces share the interesting properties of Jordan curves (simple closed curves). In the current section we show that Jordan surfaces help us to give alternative characterizations of (strong) Jordan graphs. Before getting into that, we recall some relevant results and discuss some basic properties of Jordan surfaces and of boundaries in (strong) Jordan graphs.

In Figure 3.3.1 we demonstrated that near-Jordanness of a surface does not guarantee the connectedness either of its interior or of its exterior. This changes essentially when we replace "near-Jordan" by "Jordan." The following are immediate consequences of the definition of a spel-adjacency, of Corollary 3.3.6(iii), and of Lemma 5.3.1.

Corollary 7.2.1. *If S is a Jordan surface and κ and λ are spel-adjacencies in a digital space (V, π), then S is $\kappa\lambda$-Jordan.*

Corollary 7.2.2. *If S is a Jordan surface and κ and λ are spel-adjacencies in a digital space (V, π), then there exists a binary picture (V, π, f) in which S is the unique $\kappa\lambda$-boundary.*

We may ask whether or not the converse of either Corollary 7.2.1 or Corollary 7.2.2 holds. This is not to the case in general, as illustrated by the following example in (Z^2, ω_2).

0	0	0	0	0	0	0	0	0	0	0	0	0	0	0	0	0	0
0	0	0	0	0	0	0	0	0	0	0	0	0	0	0	0	0	0
0	0	0	0	0	0	1	1	1	1	1	1	0	0	0	0	0	0
0	0	0	0	0	0	1	0	0	0	0	1	0	0	0	0	0	0
0	0	0	0	0	0	1	1	1	1	1	1	0	0	0	0	0	0
0	0	0	0	0	0	0	0	0	0	0	0	0	0	0	0	0	0
0	0	0	0	0	0	0	0	0	0	0	0	0	0	0	0	0	0

Figure 7.2.1. Illustration that the converses of Corollary 7.2.1 and of Corollary 7.2.2 do not hold.

In Figure 7.2.1, there is only one ω_2-component of the set of 1-spels and only one δ-component of the 0-spels. The unique $\omega_2\delta$-boundary between them (comprising of all surfels between a 1-spel and a 0-spel) is $\omega_2\delta$-Jordan. However, it is not minimally near-Jordan since the proper subset indicated by the heavy lines in Figure 7.2.1 also forms a near-Jordan surface.

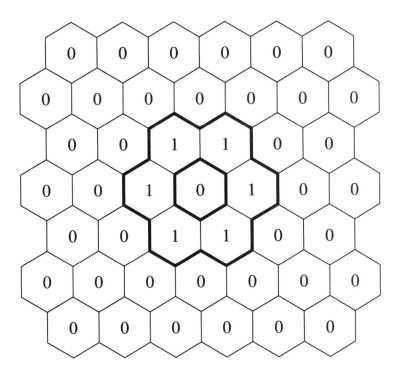

Figure 7.2.2. In the strong Jordan graph (H, ε), we define the very tight spel-adjacency ε^2 by $(c, d) \in \varepsilon^2 \Leftrightarrow d \in L^2_\varepsilon(\{c\})$. Then the unique $\varepsilon\varepsilon^2$-boundary indicated by the heavy lines in the binary picture of the figure is $\varepsilon\varepsilon^2$-Jordan, but it is not a Jordan surface.

In case the reader is wondering whether we could produce such an example only because we used a digital space which is not a strong Jordan graph and with one of the adjacencies not very tight, in Figure 7.2.2 we give an alternative example satisfying these more stringent conditions. Nevertheless, the following are immediate consequence of Theorem 5.3.8, which provides a partial converse (for the case when both κ and λ are the proto-adjacency) to the corollaries above.

Corollary 7.2.3.   *Let S be a surface in a digital space (V, π). The following are equivalent:*

   (i) *S is a near-Jordan $\pi\pi$-boundary in some binary picture over (V, π).*

   (ii) *S is a Jordan surface.*

By using the definitions of a (strong) Jordan graph and of a (strong) Jordan pair of spel-adjacencies, this corollary has the following further immediate consequence.

Corollary 7.2.4.   *Let S be a surface (respectively, a finite surface) in a strong Jordan graph (respectively, a Jordan graph) (V, π). The following are equivalent:*

   (i) *S is a $\pi\pi$-boundary in some binary picture over (V, π).*

   (ii) *S is a Jordan surface.*

Although we have presented Corollary 7.2.4 as a trivial consequence of our previous state of knowledge, we should not move on without contemplating it since it is really one of the most important results of this book. In particular, it says that if our digital space is a strong Jordan graph (and there are many such digital spaces according to the discussion following Corollary 7.1.2), then we do not have to worry whether or not a $\pi\pi$-boundary is a Jordan surface: it always is. Conversely, every Jordan surface is a $\pi\pi$-boundary. Analogous statements can be made for finite surfaces in Jordan graphs.

The question arises whether or not we could strengthen this result, i.e., is it the case that whenever S is a $\kappa\lambda$-boundary in some binary picture over (V, π) for some tight spel-adjacencies κ and λ in (V, π), then S is a Jordan surface? (It follows from Corollary 7.1.1 that S is a near-Jordan surface.) Unfortunately, the answer is no. In Figure 7.2.2 we illustrate that, even for very tight spel-adjacencies κ and λ, it is possible to have a finite $\kappa\lambda$-boundary in a finite binary picture over a strong Jordan graph which is not a Jordan surface.

Now we give the promised alternative characterizations of strong Jordan graphs and of Jordan graphs, which use the notion of Jordan surfaces.

Theorem 7.2.5.   *Let (V, π) be a digital space. The following are equivalent:*

   (i)   *(V, π) is a strong Jordan graph.*

   (ii) *The immediate interior of every $\pi\pi$-boundary is π-connected.*

   (iii) *The immediate interior of every Jordan surface is π-connected.*

   (iv) *The immediate exterior of every $\pi\pi$-boundary is π-connected.*

   (v)   *The immediate exterior of every Jordan surface is π-connected.*

Proof. We prove only the equivalence of (i), (ii), and (iii). The rest of the proof is clearly strictly analogous.

To show that (i) implies (ii), we assume that (V, π) is a strong Jordan graph. Let S be a $\pi\pi$-boundary. By Corollary 7.2.4, S is a Jordan surface and so has all the properties listed in Corollary 3.3.6. Define a binary picture (V, π, f) so that the set of 1-spels is exactly the immediate interior of S. In this binary picture, $E(S)$ is a π-connected set by Corollary 3.3.6(iii) and is in fact a π-component of the set of 0-spels of (V, π, f). This follows since, according to Corollary 3.3.6(ii), the complement in V of $E(S)$ is $I(S)$, but any element of $I(S)$ which is proto-adjacent to an element of $E(S)$ is in $II(S)$, by Corollary 3.3.6(i), and therefore is a 1-spel in (V, π, f). Now we suppose that $II(S)$ is not π-connected and show that this leads to a contradiction. Let c and d be two spels in different π-components of $II(S)$. Since $I(S)$ is π-connected, there is a π-path from c to d entirely in $I(S)$. This path must begin in one π-component J of $II(S)$ (which is also a π-component of the set of 1-spels of (V, π, f)) and must eventually arrive at another π-component L of $II(S)$ (which is also a π-component of the set of 1-spels of (V, π, f)), having gone through a π-component K (different from $E(S)$) of the 0-spels of (V, π, f). Thus, in the $\pi\pi$-adjacency graph of (V, π, f), there is a cycle $\langle E(S), J, K, \cdots, L, E(S) \rangle$ which by Corollary 7.1.6 contradicts the fact that (V, π) is a strong Jordan graph.

That (ii) implies (iii) follows immediately from Corollary 7.2.3, according to which a Jordan surface is a $\pi\pi$-boundary.

To show that (iii) implies (i), we assume that the $\pi\pi$-adjacency graph of some binary picture (V, π, f) has a cycle of the form $\langle Q, O, \cdots, O', Q \rangle$ (where Q is some fixed π-component of the set of 0-spels and O and O' are two different π-components of the 1-spels), and based on this assumption, we produce a Jordan surface whose immediate interior is not π-connected. We define a subset of the set of all π-components of the 1-spels by

$$\mathcal{O} = \{ O \mid \partial(O, Q) \text{ is not empty} \}. \tag{7.2.1}$$

We define a binary relation $\sim$ on elements of $\mathcal{O}$ by $O \sim O'$ if, and only if, either $O = O'$ or there is a cycle $\langle Q, O, \cdots, O', Q \rangle$ in the $\pi\pi$-adjacency graph of (V, π, f). It is easy to show that $\sim$ is an equivalence relation. According to our assumption, there is at least one equivalence class which contains more than one element; let us refer to the union of all the elements of this class as P. Next we show that $\partial(P, Q)$ is near-Jordan. Let $\langle c^{(0)}, \cdots, c^{(K)} \rangle$ be a π-path from $II(\partial(P, Q))$ to $IE(\partial(P, Q))$. Let $c^{(0)}$ be in O, and let $c^{(k)}$ be the last element in the π-path which is not in Q. Then $c^{(k)} \in O'$ for some O' in $\mathcal{O}$. If $O' \neq O$, then the π-path can be used to create a cycle $\langle Q, O, \cdots, O', Q \rangle$ in the $\pi\pi$-adjacency graph of (V, π, f). This shows that $O \sim O'$ and so $(c^{(k)}, c^{(k+1)}) \in \partial(P, Q)$. Now we show that $\partial(P, Q)$ is minimally near-Jordan. Let S be any surface that is a proper subset of $\partial(P, Q)$, and suppose that $(c, d) \in \partial(P, Q)$, $(c, d) \notin S$, and $(c', d') \in S$. Since $c' \in O'$ and $c \in O$ for some O and O' such that $O \sim O'$, there is a π-path from c' to c not crossing $\partial(P, Q)$ and hence S. (This is obvious if $O = O'$, which is a π-component of the set of 1-spels. In the alternative case, we can use the "inside" of the cycle $\langle Q, O', \cdots, O, Q \rangle$ to get from c' to c without crossing $\partial(P, Q)$.) By our choice of (c, d), we can extend the π-path to d without crossing S. Finally, we can extend it to d' without crossing $\partial(P, Q)$ since both d and d' are in Q, which is a π-component of the set of 0-spels. This shows that S is not near-Jordan. To complete the proof, let O and O' be two distinct elements of $\mathcal{O}$ which are subsets of P. Since $II(\partial(P, Q))$ contains at least one element from both O and O', it cannot be π-connected (otherwise O and O' would be the same π-component of the set of 1-spels). $\square$

Theorem 7.2.6.   *Let* (V, π) *be a digital space such that* π *is finitary. The following are equivalent:*

    (i) (V, π) *is a Jordan graph.*

    (ii) *The immediate interior of every finite* $\pi\pi$-*boundary is* π-*connected.*

    (iii) *The immediate interior of every finite Jordan surface is* π-*connected.*

    (iv) *The immediate exterior of every finite* $\pi\pi$-*boundary is* π-*connected.*

    (v) *The immediate exterior of every finite Jordan surface is* π-*connected.*

Proof. The proof of Theorem 7.2.5 can be adapted with a very minor change: we need to use Corollary 7.1.8 instead of Corollary 7.1.6. $\square$

A significant aspect of the last two theorems is worth emphasizing: in both of them (i), (ii), and (iv) deal with concepts defined in terms of binary pictures; however, the equivalent conditions (iii) and (v) do not in any way refer to assignment of 0s and 1s to spels, they are just properties of certain types of surfaces. In other words, we could have equivalently defined when a digital space is a (strong) Jordan graph without ever referring to binary pictures. Thus, being a (strong) Jordan graph is a purely geometrical property of a digital space; it does not require a mention of pictures in its definition. Given the path we have taken in the book — defining (strong) Jordan graphs in terms of (strong) Jordan pairs, which in turn were defined using the notion of boundaries in binary pictures — this is quite a surprising turn of events.

As an example of this idea, we see that the surface indicated by heavy lines in Figure 7.2.1 by itself is sufficient to tell us that (Z^2, ω_2) is not a Jordan graph (even if there were no 1s and 0s in that figure). This is because the surface oriented from the inside to the outside of this rectangle is clearly a finite Jordan surface, but its immediate exterior is not ω_2-connected (since the "corner" 0-spels which touch the vertices of the rectangle are not in the immediate exterior). Hence condition (v) of Theorem 7.2.6 is not satisfied, implying that (Z^2, ω_2) is not a Jordan graph. The same argument fails for the finite Jordan surfaces in Figure 7.2.2. If we consider either of the two $\varepsilon\varepsilon$-boundaries between the only ε-component of 1-spels and one of the two ε-components of the 0-spels (both of these $\varepsilon\varepsilon$-boundaries are clearly finite Jordan surfaces), then we see that the immediate exterior of the surface in question is clearly ε-connected. This is just as well since we know that (H, ε) is a Jordan graph.

We complete this section with yet another way of characterizing (strong) Jordan graphs, one which pushes the idea brought up in the last paragraphs even further: in these characterizations we do not even need the notion of a surface. For any digital space (V, π), for any spel c, and for any nonempty set D of spels not containing c, a set U of spels is called cD-*separating* if c is not in U and, for every spel d in D, every π-path from c to d contains an element of U. A cD-separating set is called *minimally* cD-*separating*, if no proper subset of it is cD-separating. It turns out that this new concept is not so different from those already introduced, as can be seen from the following.

Theorem 7.2.7.   *Let* (V, π) *be a digital space and* U *be a subset of* V. *The following are equivalent:*

    (i) U *is a minimally* cD-*separating set for some spel* c *and a* π-*connected set of spels* D.

    (ii) U *is the immediate exterior of a Jordan surface.*

Proof. First assume that U is a minimally cD-separating set for some spel c and a π-connected set D. We define a binary picture (V, π, f) so that a spel is given the value 1 if, and only if, there is a π-path from c to it which does not contain an element of U.

Clearly, c itself is a 1-spel, whereas all elements of D are 0-spels. Also, the set O of 1-spels is π-connected and all π-components of the set of 0-spels are π-components of the complement of O. Let Q be the π-component of the set of 0-spels which contains D (since D is π-connected, there is such a π-component), and let S be the $\pi\pi$-boundary $\partial(O, Q)$ (necessarily nonempty). By Theorem 3.3.3, S is $\pi\pi$-Jordan; in other words, S is a Jordan surface. Clearly, the immediate exterior of S is a subset of U. In fact, it is the whole of U since it is a cD-separating set and U is minimally cD-separating.

Now assume that U is the immediate exterior of a Jordan surface S. By Corollary 3.3.6(iii), both $I(S)$ and $E(S)$ are π-connected. Let c be any element in $I(S)$, and let $D = E(S)$. By Corollary 3.3.6(iv), $IE(S)$ is a cD-separating set. Let d be any element of $IE(S)$. Then d is also in D. Let e be an element of $I(S)$ which is π-adjacent to d. There is a π-path from c to e entirely in $I(S)$. It follows that any subset of $IE(S)$ which does not contain d would not be cD-separating. $\square$

This, combined with Theorems 7.2.5 and 7.2.6, immediately yields the following new characterizations of strong Jordan graphs and Jordan graphs. Note that these characterizations do not use the concept of a surface, but they use only the concepts of π-connected and of cD-separating sets of spels.

Corollary 7.2.8. *Let (V, π) be a digital space. The following are equivalent:*

(i) *(V, π) is a strong Jordan graph.*
(ii) *For every spel c and for every π-connected nonempty set D of spels not containing c, every minimally cD-separating set is π-connected.*

Corollary 7.2.9. *Let (V, π) be a digital space such that π is finitary. The following are equivalent:*

(i) *(V, π) is a Jordan graph.*
(ii) *For every spel c and for every π-connected nonempty set D of spels not containing c, every finite minimally cD-separating set is π-connected.*

7.3. Spel-Manifolds

We wish to complete this chapter with a little diversion, one that takes us away from the spirit of this book but introduces a quite popular alternative approach. Our philosophy has been that in a digital space an "object" is a union of spels and a "surface" is a union of surfels, which are oriented faces of the spatial elements (the spels). Thus, intuitively, we see that what we call a "surface" is made up, at least locally, of units which look like bits of surfaces in the classical sense. We have also seen that, quite often, these local surface patches (surfels) can be joined together so that they form a surface in the classical sense. Much more of this will be seen in the next chapter.

An alternative way of approaching things is the following: let us try to simplify matters by having only basic units which are of the same kind; in three-dimensional space these would be the voxels which are the Voronoi neighborhoods of the grid points. If we adopt such an approach, the question arises as to what kind of collection of voxels corresponds to a particular notion of classical geometry, for example, to the notion of a "surface." Once we have adopted a definition, there is a secondary problem: given a classical surface in R^3, can we find a digital surface that "corresponds to it" in a reasonable sense? This last problem is referred to as the problem of "surface voxelization." Since the concepts introduced in this section are not used elsewhere in the book, the reader may move on to the next chapter without any danger of losing the thread of our presentation.

We concentrate on the approach to surface voxelization advocated in [6]. First we need to discuss what kind of surfaces $F \subset R^3$ we wish to voxelize. For this intuitive discussion, we use mathematical terms loosely (without carefully defining them). We will be more careful later, when we will try to prove some precise results. Basically, we are interested in surfaces F which are 2-*manifolds*. This means, that for any point $c \in F$, there exists a ball $B_{r,c}$ (of some radius r whose value may vary with c), such that $B_{r,c} \cap F$ is topologically equivalent (in the sense of being continuously deformable) to a disk. For the voxelization to meaningfully capture the appearance of the surface, we need to be more restrictive than this: we can allow only surfaces in which the radii in the previous sentence are bounded below by a fixed positive real number. (The size of this real number depends on the grid that is used for the voxelization. Clearly, we cannot hope to capture details in surfaces which are small compared to the spacing of the grid points.) $B_{r,c} - F$ has two components; in other words, the neighborhood of a point in the surface is separated by the surface. (The notation $A - B$ denotes the *set-theoretical difference*, i.e., the set of all those elements of A which are not also in B.)

When we come to putting conditions on what we consider a sound voxelization of a surface, we need to refer to notions such as adjacency, component, path, and neighborhood in the digital space. All of these are defined on the basis of various spel-adjacencies. For example, the ρ-neighborhood of a spel is the set of all spels ρ-adjacent to it. A basic problem is that in some digital spaces, such as (Z^3, ω_3), one has to use different adjacencies in different places to end up with a consistent set of definitions [6]. This is where the notion of a Jordan graph promises to become useful: using it we may be able to avoid having to deal simultaneously with a whole zoo of spel-adjacencies. With all this in mind, we look at the general conditions of [6] for considering a set P^F of spels in a digital space (V, π) as a voxelization of a surface F in R^3. The first three conditions are quite general. In fact they are meaningful in any digital space, and they do not refer to the notion of a surface in R^3. Between them they provide a digital version of the topological notion of a manifold.

In any digital space (V, π), a nonempty subset P of V is called a *spel-manifold* if it satisfies the following three conditions.

(i) P is π-connected.

(ii) For each $c \in P$, $R_\pi(\{c\}) - P$ has two π-components.

(iii) For each $c \in P$ and for each $d \in R_\pi(\{c\}) \cap P$, $R_\pi(\{d\})$ has a nonempty intersection with both π-components of $R_\pi(\{c\}) - P$.

An example of a spel-manifold in (F, β_1), the fcc grid with face-adjacency, is given by $P = \{c \in F \mid c_3 = 0\}$. For all spels $(c_1, c_2, 0)$ in P, $c_1 + c_2$ is even. By the definition of β_ϕ, two spels c and d in P are β_1-adjacent if, and only if, $|c_1 - d_1| = |c_2 - d_2| = 1$. To see that (i) is satisfied, let c and d be arbitrary elements in P. They are β_1-connected in P by the β_1-path (see Figure 7.3.1) that starts at $c = (c_1, c_2, 0)$, goes through $(d_2 + c_1 - c_2, d_2, 0)$ to $((d_1 + d_2 + c_1 - c_2)/2, (d_1 + d_2 - c_1 + c_2)/2, 0)$ — that the components of this vector are integers and that it belongs to P follows from the already stated properties of the elements of P — and then to $d = (d_1, d_2, 0)$. To see that

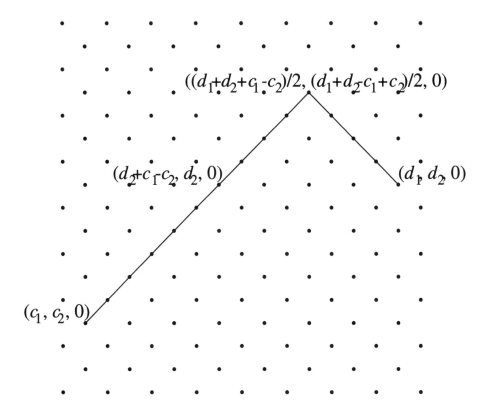

Figure 7.3.1. Illustration that the subset of the fcc grid defined by $c_3 = 0$ is β_1-connected.

(ii) is satisfied, consider all the spels in $F - P$ which are proto-adjacent to an arbitrary $c = (c_1, c_2, 0)$ in P. They are $(c_1 + 1, c_2, 1)$, $(c_1, c_2 + 1, 1)$, $(c_1 - 1, c_2, 1)$, $(c_1, c_2 - 1, 1)$ and $(c_1 + 1, c_2, -1)$, $(c_1, c_2 + 1, -1)$, $(c_1 - 1, c_2, -1)$, $(c_1, c_2 - 1, -1)$. The first four are β_1-connected (the order in which they are given is in fact a β_1-path) and so are the second four. On the other hand, none of the first four is proto-adjacent to any of the second four since the norm of the difference of a spel in the first four and a spel in the second four has to be at least 2. To verify (iii), consider an arbitrary $c = (c_1, c_2, 0)$ in P and a $d = (d_1, d_2, 0)$ in P which is proto-adjacent to c. Then, either $d_1 = c_1 + 1$, $d_2 = c_2 + 1$ or $d_1 = c_1 + 1$, $d_2 = c_2 - 1$ or $d_1 = c_1 - 1$, $d_2 = c_2 + 1$ or $d_1 = c_1 - 1$, $d_2 = c_2 - 1$. In the first case, d is proto-adjacent to both $(c_1 + 1, c_2, 1)$ and to $(c_1 + 1, c_2, -1)$, which are in different β_1-components of $R_{\beta_1}(\{c\}) - P$, as we have just seen while verifying (ii). Each of the other three cases can be handled similarly.

As an interesting contrast to this example, consider the set $P = \{ c \in Z^3 \mid c_3 = 0 \}$ of spels in the digital space (Z^3, ω_3); i.e., in the cubic grid with face-adjacency. We leave it to the reader to provide the easy proof of the fact that P is ω_3-connected. For each $c \in P$, $R_{\omega_3}(\{c\}) - P$ has only two elements, $(c_1, c_2, 1)$ and $(c_1, c_2, -1)$, which are not proto-adjacent. On the other hand, they are also not proto-adjacent to any element of P other than c. It follows that whereas P satisfies the first two conditions in the definition of a spel-manifold, it fails the third. On the other hand, one has the intuitive conviction that P should surely be considered a valid voxelization of the continuous surface $F = \{v \in R^3 \,|\, v_3 = 0\}$. It is for such reasons that in the approach of [6] spel-adjacencies in addition to the proto-adjacency are brought into the discussion of what is a valid voxelization in (Z^3, ω_3) of a surface in R^3. Here we look into the possibility of avoiding having to do this by considering spel-manifolds in Jordan graphs. Our first result is, in fact, quite general.

Theorem 7.3.1.   *If P is a spel-manifold in a digital space (V, π), then $V - P$ has at most two π-components.*

Proof. First we show that any spel c in $V - P$ is π-connected in $V - P$ to some spel in $R_\pi(P) - P$. Since P is a nonempty subset of V and V is π-connected, there is a π-path $\langle c^{(0)}, \cdots, c^{(K)} \rangle$ in V connecting c to an element of P. Let k be the smallest integer such that $c^{(k)} \in P$. We know that $k > 0$, and so $\langle c^{(0)}, \cdots, c^{(k-1)} \rangle$ is a π-path in $V - P$ connecting c to an element of $R_\pi(P) - P$.

In view of this, now we can complete the proof by showing that $R_\pi(P) - P$ has at most two π-components. For this purpose, let c be an arbitrary element of P, and let d and e be elements of the two π-components of $R_\pi(\{c\}) - P$. These exist by property (ii) of a spel-manifold. For an arbitrary element f of $R_\pi(P) - P$, there exists a π-path $\langle c = c^{(0)}, \cdots, c^{(K)} \rangle$ in P, such that $f \in R_\pi(\{c^{(K)}\}) - P$ (this is by property (i) of a spel-manifold). We prove by induction on K that f is π-connected in $R_\pi(P) - P$ to either d or e. If $K = 0$, then by property (ii) of a spel-manifold f is π-connected in $R_\pi(\{c\}) - P$ (and hence in $R_\pi(P) - P$) to either d or e. Now suppose that $K > 0$ and that the result holds for $K - 1$. By property (iii) of a spel-manifold, $R_\pi(\{c^{(K-1)}\})$ has a nonempty intersection with both the components of $R_\pi(\{c^{(K)}\}) - P$. This means that there is a $g \in R_\pi(\{c^{(K-1)}\}) - P$ which is π-connected in $R_\pi(\{c^{(K)}\}) - P$ (and hence in $R_\pi(P) - P$) to f. By the induction hypothesis, g is π-connected in $R_\pi(P) - P$ to either d or e, and the result follows. $\square$

The proof of this theorem has made essential use of all three properties of a spel-manifold, but made no other assumptions; it is applicable to any digital space whatsoever. It would be very satisfactory if we could now prove that $V - P$ has exactly two π-components. Unfortunately this is not necessarily true even if we restrict our attention to 1-simply connected digital spaces.

Theorem 7.3.2.   *There exists a 1-simply connected digital space (V, π) and a spel-manifold P in it such that $V - P$ has only one π-component.*

Proof. Consider Figure 7.3.2. The digital space represented on the left is 1-simply connected. This is easy to see since any path of length four or greater is elementarily 1-equivalent to a shorter path. ($\langle A, C, a, B \rangle$ is elementarily 1-equivalent to $\langle A, C, B \rangle$, $\langle A, C, b, A \rangle$ is elementarily 1-equivalent to $\langle A, C, A \rangle$, $\langle A, C, B, A \rangle$ is elementarily 1-equivalent to $\langle A, C, A \rangle$, $\langle C, a, B, c \rangle$ is elementarily 1-equivalent to $\langle C, B, c \rangle$, etc.)

The digital space (V, π) represented on the right of Figure 7.3.2 is obtained from the digital space which is represented on the left by the addition of some extra surfels. However, the proto-adjacency π of the digital space represented on the right is a 2-limited spel-adjacency in the digital space represented on the left (for example, for $(A, b) \in \pi$, there is the path $\langle A, C, b \rangle$ of length 2 in the digital space on the left) and so, by Theorem 6.1.1, (V, π) is also 1-simply connected.

Clearly, $P = \{A, B, C\}$ is π-connected. Further, $R_\pi(\{A\}) - \{P\}$ has two π-components $\{B, c\}$ and $\{C, b\}$, $R_\pi(\{B\}) - \{P\}$ has two π-components $\{C, a\}$ and $\{A, c\}$, and

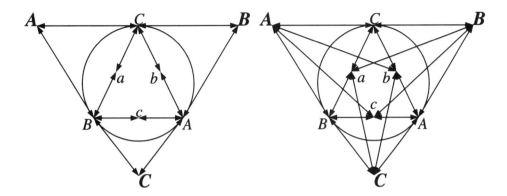

Figure 7.3.2. On the left we represent a digital space which is easily seen to be 1-simply connected. On the right we introduce some additional proto-adjacencies. The resulting space (V, π) is still 1-simply connected. In this space $P = \{A, B, C\}$ is a spel-manifold and $V - P$ has only one π-component.

$R_\pi(\{C\}) - \{P\}$ has two π-components $\{A, b\}$ and $\{B, a\}$. Thus we see that all conditions for a spel-manifold are fulfilled by P. Finally, it is easy to see that $V - P$ is π-connected. $\square$

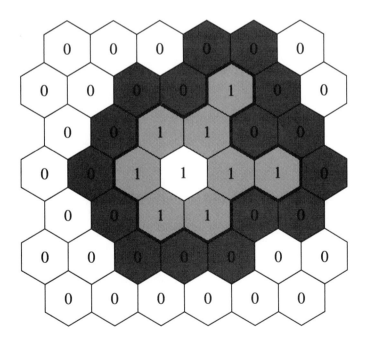

Figure 7.3.3. In the strong Jordan graph (H, ε), we define a finite picture with only one $\varepsilon\varepsilon$-boundary S (indicated by the heavy lines). Note that both the lightly shaded $II(S)$ and the heavily shaded $IE(S)$ fail to satisfy condition (ii) of the definition of a spel-manifold.

Thus spel-manifolds, as defined above, fail to have a desirable "Jordan" property (the complement of the spel-manifold does not necessarily have exactly two components), even if we restrict our attention to Jordan graphs. Although there are certainly alternative approaches in the literature to achieving something of this sort, we do not pursue this any further in this book. Since our basic approach of defining surfaces as a collection of surfels provides us with all the necessary Jordan properties in a very pleasing way, we will now abandon this diversion to spel-manifolds and concentrate on continuing with the discussion of such surfel-based surfaces. For those readers who persist in their desire to represent surfaces by sets of pixels, all we have to suggest is the use of the immediate interior or of the immediate exterior (or even of the immediate neighborhood) of a Jordan surface. In a strong Jordan graph these are necessarily connected (see Theorem 7.2.5), but they are not necessarily spel-manifolds, as can be seen from Figure 7.3.3.

7.4. Exercises

7.1.   Show that every bridged digital space (as defined in Exercise 6.16) is a strong Jordan graph.

7.2.   Define the *arc set associated with a surface* S in a digital space as the set

$$A(S) = \{ \, \{c, d\} \mid (c, d) \in S \text{ or } (d, c) \in S \, \} \tag{7.4.1}$$

of unordered pairs of spels and the *graph associated with a digital space* (V, π) as

$$G(V, \pi) = (V, A(\pi)). \tag{7.4.2}$$

Let a nonempty subset F of $A(\pi)$ be called a *cut* in the graph $G(V, \pi)$, if there exists a subset O of V such that

$$F = A\big(\partial(O, \overline{O})\big). \tag{7.4.3}$$

Prove the following.
  (i) If a surface S in a digital space (V, π) is near-Jordan, then $A(S)$ is a cut in the graph $G(V, \pi)$. If a nonempty subset F of $A(\pi)$ is a cut in the graph $G(V, \pi)$, then there is a near-Jordan surface S in (V, π) such that $A(S) = F$.
  (ii) A digital space (V, π) is a strong Jordan graph (respectively, a Jordan graph) if, and only if, for every binary picture (V, π, f) and for every pair of π-components O of the 1-spels and Q of the 0-spels such that $\partial(O, Q)$ is nonempty (respectively, nonempty and finite), $A(\partial(O, Q))$ is a cut in the graph $G(V, \pi)$.

7.3.   As done in Figure 7.1.2 for Figure 7.1.1, draw
  (i) the $\nu_4 \nu_1$-adjacency graph for Figure 5.3.1;
  (ii) the $\beta\gamma$-adjacency graph for Figure 5.3.2;
  (iii) the $\nu\nu$-adjacency graph for Figure 5.4.1;
  (iv) the $\omega_2 \delta$-adjacency graphs for Figures 5.4.2 and 5.4.4;
  (v) and, for your definition of ρ the $\rho\omega_2$-adjacency graph for Figure 5.6.1.

Which of these graphs are acyclic?

7.4. For the ϕ defined on page 42, draw the $\phi\omega_4$-adjacency graph of the binary picture (Z^4, ω_4, f), where $f(c) = 1$ if, and only if, $-1 \leq c_i \leq 1$ for $1 \leq i \leq 4$ and $\sum_{i=1}^{4} c_i$ is odd.

7.5. Give a detailed proof of Theorem 7.2.6.

7.6. Show that $P = \{c \in Z^2 \,|\, c_1 = c_2\}$ is a spel-manifold in (Z^2, β), but it is not a spel-manifold in (Z^2, γ) or in (Z^2, α_2).

7.7. Show that $P = \{c \in Z^3 \,|\, c_1 = c_2\}$ is not a spel-manifold in (Z^3, δ_3).

7.8. Show that $P = \{c \in F \,|\, c_1 = c_2\}$ is not a spel-manifold in (F, β_1).

7.9. Is $P = \{c \,|\, c_4 = 0\}$ a spel-manifold in (Z^4, ϕ)? (See page 42 for the definition of ϕ.)

7.10. Show that in Figure 7.1.3 $P = \{3, 4\}$ is a spel-manifold in the strong Jordan graph (V, π) described by Figure 7.1.3 and yet $V - P$ has only one π-component.

7.11. Let S be a surface in a digital space (V, π). Show that $IN(S)$ is π–connected, provided that at least one of the following two conditions hold.
    (i) (V, π) is a strong Jordan graph and S is a Jordan surface.
    (ii) (V, π) is a Jordan graph and S is a finite Jordan surface.

8
Boundary Tracking

"Vezess, vezess, uj célra, Lucifer
Ugy is sok szép időt vesztettem el
Ez ál-uton."
("Lead me, lead me, to a new aim, Lucifer
As it is I have wasted plenty of good time
On this false track.")

I. Madách, *Az Ember Tragédiája* (*The Tragedy of Man*), Scene IV.

8.1. Tracking in Finitary 1-Simply Connected Spaces

Now we are ready to start proving rigorously that boundary tracking algorithms perform exactly as they are intended to do. As an initial demonstration of this, we show that there is a "one-size-fits-all" algorithm which, given a binary picture over a finitary 1-simply connected digital space and a boundary face between a 1-spel and a 0-spel, will return the set of all boundary faces between the component of 1-spels containing the given 1-spel and the component of 0-spels containing the given 0-spel, provided only that this set is finite. We show that a proof of correctness of this algorithm is an immediate consequence of some the general results presented in the earlier chapters.

We propose an algorithm for the following task:
Given a binary picture (V, π, f) over a finitary 1-simply connected digital space (V, π), a 1-spel c, and a 0-spel d such that $(c, d) \in \pi$ and the boundary $\partial(O, Q)$ between the π-component O of the set of 1-spels containing c and the π-component Q of the set of 0-spels containing d is finite,
find $\partial(O, Q)$.

We present the algorithm in a manner which resembles as closely as possible Artzy's Algorithm, which was stated in Chapter 1 and whose correctness (in the sense of Claim 1.11.2) is proved below in the current chapter. We are now a bit more formal, however, than we were in Chapter 1 and give names (L and S, respectively) to the queue used by the algorithm and to its output. Putting a surfel into S corresponds to the notion of "dirtying a face" in Artzy's Algorithm.

1-Simply Connected Algorithm
(1) Put (c, d) into L and S.
(2) Remove a surfel (e, g) from L.
   a.   For all h such that $f(h) = 0$, $(h, g) \in \pi$, $(e, h) \in \pi - S$,
       put (e, h) into L and S.
   b.   For all h such that $f(h) = 1$, $(e, h) \in \pi$, $(h, g) \in \pi - S$,
       put (h, g) into L and S.
(3) Check if L is empty.
   a.   If it is, STOP.
   b.   If it is not, start again at Instruction (2).

A simple example is provided by Figure 8.1.1 with $O = \{o\}$, $Q = \{a, b, c, d, m, n, \cdots\}$. With the specification of (o, d) as the starting surfel, the algorithm might proceed as follows.

	L	S
	$\{(o,d)\}$	$\{(o,d),$
(o,d)	$\{(o,c), (o,m)\}$	$(o,c), (o,m),$
(o,c)	$\{(o,m), (o,b)\}$	$(o,b),$
(o,m)	$\{(o,b), (o,n)\}$	$(o,n),$
(o,b)	$\{(o,n), (o,a)\}$	$(o,a)\}$
(o,n)	$\{(o,a)\}$	
(o,a)	$\{ \}$	

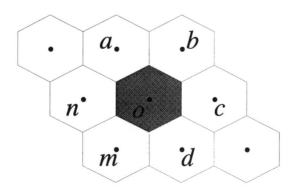

Figure 8.1.1. The picture used in the simple example to illustrate the 1-Simply Connected Algorithm.

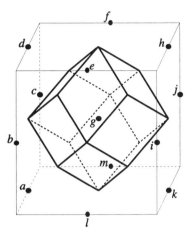

Figure 8.1.2. An element g of the face-centered cubic grid and the twelve elements adjacent to it.

The behavior of the algorithm is much more complicated when applied to the digital space based on the face-centered cubic grid. In that case L can become very large, but (as we will see below) it is emptied eventually. For example, if in Figure 8.1.2 g is a 1-spel and all its neighbors are 0-spels, then with the specification of (g, d) as the starting surfel, the first application of Instruction (2)a will put all of (g, e), (g, f), (g, b) and (g, c) into both L and S.

To see how the validity of the algorithm depends on the 1-simply connectedness of the underlying digital space, we consider its behavior on a very simple finitary 2-simply connected digital space. (This space is the simplest digital equivalent of a circle, which is not a simply connected object in classical topology.) Consider Figure 8.1.3 in which proto-adjacency between spels is defined by sharing exactly one whole edge (thus a and c are not proto-adjacent) and the only 1-spel is c. Specifying (c, d) as the starting surfel, the algorithm terminates with $S = \{(c, d)\}$, i.e., it does not output the element $(c, b) \in \partial(\{c\}, \{a, b, d\})$.

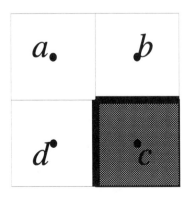

Figure 8.1.3. Illustration of a finitary 2-simply connected digital space (with only four spels) in which the 1-Simply Connected Algorithm fails to detect a boundary.

Now we prove the main result of this section, which is that the 1-Simply Connected Algorithm given above terminates and that at that time $S = \partial(O, Q)$. We do this through a series of lemmas, the first of which does not need the finitariness of the underlying digital space or the finiteness of the boundary being detected.

Lemma 8.1.1. *Both just before and just after the execution of Instruction (2) in the 1-Simply Connected Algorithm, S is a surface and $S \subset \partial(O, Q)$.*

Proof. After the execution of Instruction (1) in the 1-Simply Connected Algorithm, $S = \{(c, d)\}$, and so it is a surface which is a subset of $\partial(O, Q)$. Note also that at this time $L = \{(c, d)\} \subset \partial(O, Q)$. To complete the proof of the lemma, now we show that if just before the execution of Instruction (2) S is a surface, $S \subset \partial(O, Q)$ and $L \subset \partial(O, Q)$, then the same is true at any time during (and, in particular, just after) the execution of Instruction (2). Since S is not changed due to Instruction (3), this is sufficient to prove the lemma.

For the (e, g) that we remove from L at the beginning of the execution of Instruction (2), $e \in O$ and $g \in Q$. An (e, h) is put into L and S only if $f(h) = 0$, $(h, g) \in \pi$ and $(e, h) \in \pi$, i.e., only if $h \in Q$ and $(e, h) \in \partial(O, Q)$. A similar argument holds for the (h, g) that may be put into L and S. Since nothing is ever removed from S, it remains a surface. $\square$

Lemma 8.1.2. *If (V, π) is finitary and $\partial(O, Q)$ is finite, then the 1-Simply Connected Algorithm terminates in a finite number of steps.*

Proof. The finitariness of π implies that there can be only finitely many h satisfying the conditions in either Instruction (2)a or in Instruction (2)b, and so the algorithm cannot get stuck in executing Instruction (2). The length of L is increased only when that of S is also increased. By the finiteness of $\partial(O, Q)$ and Lemma 8.1.1, this can happen only a finite number of times. Since the length of L is decreased at the beginning of Instruction (2) and can be increased only a finite number of times, sooner or later L will become empty, and the algorithm will terminate. $\square$

Lemma 8.1.3. *If the 1-Simply Connected Algorithm terminates, then at that time S is 1-locally-Jordan.*

Proof. Suppose that the 1-Simply Connected Algorithm has terminated and hence L is empty. Let $\langle e, h, g \rangle$ be a π-path such that $(e, g) \in S$. Since all things that have been put into S have also been put into L at the same time (and hence have been eventually removed from L), consider the situation when (e, g) has just been removed from L. Suppose that $f(h) = 0$. Since $(h, g) \in \pi$ and $(e, h) \in \pi$, (e, h) would be put into S at this time if it is not in S already. On the other hand, (h, g) is never put into S (this follows from Lemma 8.1.1, noting that both h and g are 0-spels). So exactly one of (e, h) and (h, g) is in S. If $f(h) = 1$, the proof is similar. $\square$

Lemma 8.1.4. *If the 1-Simply Connected Algorithm terminates, then at that time S is near-Jordan.*

Proof. Suppose that the 1-Simply Connected Algorithm has terminated. Look at the conditions of Lemma 6.2.4. S is 1-locally-Jordan, by the previous lemma. S is antisymmetric as a consequence of Lemma 8.1.1. The underlying digital space is assumed to be 1-simply connected. The way the algorithm builds S assures that $II(S)$ is π-connected, and so for any two spels in it there is a π-path connecting them entirely in $II(S)$. The crossing parity of this π-path through S has to be zero since all elements in the π-path are 1-spels. All conditions of Lemma 6.2.4 are satisfied, and so S is near-Jordan. $\square$

Theorem 8.1.5. *If (V, π) is a finitary 1-simply connected digital space and $\partial(O, Q)$ is finite, then the 1-Simply Connected Algorithm terminates in a finite number of steps and at that time $S = \partial(O, Q)$.*

Proof. The first part of the claim of this theorem is just Lemma 8.1.2. To complete the proof, first we note that according to Corollaries 7.1.2 and 7.2.4, $\partial(O, Q)$ is a Jordan surface and hence (recall Theorem 3.3.5) a minimally near-Jordan surface. This means that any near-Jordan surface that is a subset of $\partial(O, Q)$ has to be in fact the whole of $\partial(O, Q)$. By the previous lemma, at the time of termination S is a near-Jordan surface and, by Lemma 8.1.1, S is a subset of $\partial(O, Q)$, and so the result follows. $\square$

In this proof of correctness, we have made essential use of Lemma 6.2.4, of Theorem 3.3.5, and of Corollaries 7.1.2 and 7.2.4. These are general results from the mathematical theory of digital spaces. Nevertheless, knowing them allowed us to give a brief proof of correctness of a powerful algorithm. Therefore we have provided an example to illustrate our belief that the mathematical theory of digital spaces enables us to prove relatively easily the correctness of general-purpose image-processing algorithms.

There is one implementational difficulty with the 1-Simply Connected Algorithm; it is the same as we have raised regarding the Algorithm Simulating Cloning Flies (and which has motivated the introduction of Artzy's Algorithm). This difficulty arises in Instructions (2)a and (2)b: how would we in practice go about finding out whether or not $(e, h) \in \pi - S$ and whether or not $(h, g) \in \pi - S$? (Belonging to $\pi - S$ corresponds to being a not dirty face in the Algorithm Simulating Cloning Flies.) Since in practical applications S can get very large, checking on a spel being in S (and this has to be done again and again in the 1-Simply Connected Algorithm) may require a great deal of computational resources. In Chapter 1 we introduced Artzy's Algorithm to deal with the corresponding problem. In the current chapter we generalize the idea which gives Artzy's Algorithm its efficiency, so that we can do computationally efficient boundary tracking in a variety of digital spaces.

8.2. Efficient Tracking of Boundary Elements

We introduce some new notation and terminology that will simplify our exposition. We begin with some further ideas concerning binary relations in general (see Section 1.4) and follow with the introduction of new concepts regarding binary pictures.

For any binary relation ρ on a set M, we define its *transitive closure* as the binary relation ρ^* on M, defined by: $(c, d) \in \rho^*$ if, and only if, c is ρ-connected to d. As pointed out in Section 1.4, ρ^* is always a reflexive and transitive relation on M. If, in addition, ρ^* also happens to be symmetric, then it is an equivalence relation and so partitions M into ρ-components.

If O is the set of all 1-spels and Q is the set of all 0-spels in a binary picture (V, π, f), then we refer to elements of $B = \partial(O, Q)$ as *bels* (short for boundary elements) in (V, π, f). **For the rest of this chapter, we assume that B is finite and not empty.** (We do this purely for the convenience of being able to use digraphs. Much of what we prove can be proven under a weaker assumption; see Exercise 8.15.) We further suppose that there is a digraph (B, λ) with the property that λ^* is symmetric (in this context such a λ is referred to as the *bel-adjacency*). For any b in B, we use the notation $in_\lambda(b)$ to denote the number of different elements a in B for which $(a, b) \in \lambda$. Since B is finite, $in_\lambda(b)$ is always a nonnegative integer; it is called the *indegree* of the bel b.

As a simple example, consider the pair of sugar cubes in Figure 1.8.2. In our more precise terminology, we are dealing with a binary picture (Z^3, ω_3, f) in which there are only two 1-voxels and they are ω_3-adjacent to each other. Thus there are ten bels (surfels between a 1-voxel and a 0-voxel) and they can be labeled, as in Figure 1.8.2, by *I1, I2, I3, I4, I5, II1, II2, II4, II5, II6*. The bel-adjacency λ is indicated by the arrows; for example, *I2* is λ-adjacent to *I3* and *I4*, whereas *II2* is λ-adjacent to *I2* and *II4*. It is easily seen that, in this case, $in_\lambda(b)$ is always 2.

We propose an algorithm for the following task:
Given a binary picture (V, π, f) such that the set B of bels is finite, a bel-adjacency λ, and a bel o,
find the λ-component of B which contains i.
(This λ-component is the mathematically precise version of the idea of the "boundary containing the boundary face i" discussed in Chapters 1 and 2.)

Again we present the algorithm in a manner which resembles as closely as possible Artzy's Algorithm, but giving names (L, T and S, respectively) to the queue, the list, and the output of the algorithm. Putting a bel into S corresponds to the notion of "dirtying a face" in Artzy's Algorithm.

General Bel-Tracking Algorithm

(1) Put i into L and S and put $in_\lambda(i)$ copies of i into T.

(2) Remove a bel a from L. For all bels b which are λ-adjacent from a, try to find one copy of b in T.

    a.   If successful, remove this copy of b from T.

    b.   If not, then put b into L and S and put $in_\lambda(b) - 1$ copies of b into T.

(3) Check if L is empty.

    a.   If it is, STOP.

    b.   If it is not, start again at Instruction (2).

Before proving the correctness of this algorithm, some remarks are in order. First, in Instruction (2)b, $in_\lambda(b) - 1$ is guaranteed to be nonnegative. This is because we only get to this point in the algorithm for a b which is λ-adjacent from a bel a and, so, $in_\lambda(b) \geq 1$. Second, the potential efficiency of this algorithm (as opposed to the 1-Simply Connected Algorithm) comes from the fact that we check for membership in T (rather than in S). Although S keeps getting bigger and bigger as the algorithm is executed due to Instruction (2)b, the same is not true for T: elements from T will be repeatedly removed due to Instruction (2)a. Hence the size of T is likely to be a small fraction of the size of S after the algorithm has been performing for a while on a large data set. Finally, for the binary picture and bel-adjacency represented by Figure 1.8.2, $in_\lambda(b) = 2$ for every bel b. If we start the General Bel-Tracking Algorithm with *I2* as the choice for i, then its performance (using a first-in first-out queuing discipline) will be identical to that of Artzy's Algorithm (as reported in Table 1.11.1). This is not accidental; in the next section we show that Artzy's Algorithm is always a special case of the General Bel-Tracking Algorithm.

The essence of the proof of correctness is given in the next lemma. To state it easily we use a couple of abbreviations. We let $n_T(b)$ abbreviate "the number of copies of the bel b in the list T." The other definition is more complicated. The value of $n_a(b)$ is 0 if $b \notin S$ and is "$in_\lambda(b)$ less the number of bels in $S - L$ that are λ-adjacent to the bel b" otherwise.

Lemma 8.2.1.   *Both just before and just after the execution of Instruction (2) in the General Bel-Tracking Algorithm, for every bel b,*
   (i) $n_T(b) = n_a(b)$,
   (ii) *the bel b has so far been put into L and S — either due to Instruction (1) or due to Instruction (2)b — at most once, and*
   (iii) *if the bel b is in S, then i is λ-connected in B to b.*

Proof. Consider the situation just after the execution of Instruction (1). We have that $n_T(i) = in_\lambda(i) = n_a(i)$, and bel i has so far been put into L and S exactly once. For any bel b other than i, $n_T(b) = 0 = n_a(b)$, and b has not so far been put into L and S even once. Since only i has been put into S, (iii) is clearly satisfied at this time.

Now we show that if (i), (ii), and (iii) hold just before the execution of Instruction (2), then they also hold just after its execution. This is sufficient, since the situation cannot change as a result of Instruction (3). Therefore assume that we are just at the beginning of executing Instruction (2). This means that at this time L is not empty and so we remove a bel a from it. This a must have been put into L and S earlier and, since nothing is ever removed from S, we have that $a \in S$. We leave it to the reader to supply the easy proof of the fact that for those bels which are not λ-adjacent from a, the inductive step is valid. For the bels b which are λ-adjacent from a, we study two possibilities separately.

Case a: we find a copy of b in T. In this case, by Instruction (2)a, we remove this b from T. This reduces $n_T(b)$ by 1. Since b was in T, it also had to be in S (nothing is ever put into T without being put into S at the same time). Therefore the applicable part of the definition of $n_a(b)$ is that it is $in_\lambda(b)$ less the number of bels in $S - L$ that are λ-adjacent to the bel b. The only thing that changes in this definition is that a, which is λ-adjacent to b, was removed from L. Hence $n_a(b)$ is also decreased by 1, proving (i) of the lemma for the bel b. Since nothing is put into L and S, (ii) and (iii) are also valid at the end of executing Instruction (2).

Case b: there is no copy of b in T, i.e., $n_T(b) = 0$. First we show that under these circumstances it cannot be the case that b has been previously put into L and S. This is so since otherwise before the beginning of Instruction (2), the applicable part of the definition of $n_a(b)$ would be "$in_\lambda(b)$ less the number of bels in $S - L$ that are λ-adjacent to the bel b." Since at that time the bel a is still in L, the value of $n_a(b)$ has to be positive, contradicting the truth of (i) in the induction hypothesis. Upon executing Instruction (2)b, b has been put into L and S (for the first time), and $n_T(b) = in_\lambda(b) - 1$. There is at least one bel, namely, a, that is in $S - L$ and is λ-adjacent to b. There cannot be another one since whenever a bel is put into S, it is also put into L at the same time, and so if at a later time it is no longer in L, then it must have been removed from it. At that time, b would have been put into L and S, and we would not be in Case b. This shows that in this case too, $n_T(b) = n_a(b)$ just after the execution of Instruction (2). Finally, since $a \in S$ just before the execution of Instruction (2), i is λ-connected in B to a by (iii) of the induction hypothesis. The same must be true for the new bel b in S since a is λ-adjacent to it. $\square$

Theorem 8.2.2.   *If (V, π, f) is a binary picture such that the set B of bels is finite, λ is a bel-adjacency, and i is a bel, then the General Bel-Tracking Algorithm terminates in a finite number of steps and, at that time, S is the λ-component of B which contains i.*

Proof. Termination in a finite number of steps easily follows from (ii) of the previous lemma: since each of the finitely many bels is put into L at most once (irrespective how long the algorithm is run) and in each execution of Instruction (2) of the algorithm a bel is removed from L, sooner or later L has to become empty and the algorithm will stop due to Instruction (3). At that time, as indeed all through the execution of the algorithm, i is λ-connected in B to every element of S. That the converse is also true (and hence S is the λ-component of B which contains i) can be shown as follows. For any bel b of the λ-component of B which contains i, there is a λ-path $\langle b^{(0)}, \cdots, b^{(K)} \rangle$ from i to b. It is a trivial matter to show by induction that, for $1 \leq k \leq K$, $b^{(k-1)}$ will get put into L and S and, since L gets eventually emptied, $b^{(k-1)}$ must be removed from L, resulting in $b^{(k)}$ being put into L and S (provided that it is not in T, which would imply that it has been put into L and S in some previous step). $\square$

This theorem shows that the General Bel-Tracking Algorithm is powerful stuff. Its practical usefulness depends on two properties of the bel-adjacency λ. The first is, how easy is it to compute the bels λ-adjacent from a given bel? Clearly, the efficiency of executing Instruction (2) depends on this (as well as on how easy it is to determine for a bel whether or not it is in T). The other property has to do with the usefulness of the resulting boundaries: do they correspond to the $\kappa\kappa'$-boundaries for some Jordan pair $\{\kappa, \kappa'\}$ of spel-adjacencies? In the following sections we will answer these questions positively for some specific choices of the bel-adjacency in certain digital spaces.

Before going into these specifics, we discuss the relationship of the General Bel-Tracking Algorithm to its less efficient version. This is needed, in particular, to prove the validity of some of the claims in Chapter 1. First we state a technical result.

Lemma 8.2.3. *Just before the execution of Instruction (2) in the General Bel-Tracking Algorithm, for any bel a in L and for any bel b λ-adjacent from a, $b \in S$ if, and only if, $b \in T$.*

Proof. If $b \notin S$, then (by the definition just before Lemma 8.2.1) $n_a(b) = 0$ and so, by Lemma 8.2.1(i), $n_T(b) = 0$, i.e., $b \notin T$. Now suppose that $b \in S$ and that some a in L is λ-adjacent to b. Then $n_a(b) > 0$ and so, again by Lemma 8.2.1(i), $b \in T$. $\square$

Inefficient Bel-Tracking Algorithm

(1) Put i into L and S.
(2) Remove a bel a from L. For all bels b not in S which are λ-adjacent from a, put b into L and S
(3) Check if L is empty.
    a.   If it is, STOP.
    b.   If it is not, start again at Instruction (2).

Theorem 8.2.4. *If the same queuing discipline is used for L in the General Bel-Tracking Algorithm and in the Inefficient Bel-Tracking Algorithm, then just before and just after an execution of Instruction (2) the sets L and S are the same for the two algorithms. Furthermore, one of the algorithms stops after a number of cycles through Instructions (2) and (3) if, and only if, the other algorithm stops after the same number of cycles through Instructions (2) and (3).*

Proof. Just before the first execution of Instruction (2), $L = S = \{i\}$ in both algorithms. Now assume that just before some execution of Instruction (2), $L = S$ and that a is the bel that is removed from L in both algorithms. By the previous lemma, for any bel b λ-adjacent from a, $b \in S$ if, and only if, $b \in T$ in the General Bel-Tracking Algorithm. It follows that the same bels are put into L and S during the execution of Instruction (2) of the two algorithms. The final claim of the theorem is a consequence of the fact that L becomes empty in the two algorithms after the same number of executions of Instruction (2). $\square$

8.3. Boundary Tracking on Hypercubes

The first use to which we put the General Bel-Tracking Algorithm is the tracking of boundaries in the spaces $\left(Z^N, \omega_N\right)$ with $N \geq 2$. The approach that we take here will be a generalization of the specific boundary-tracking procedures (for flat flies and for fat flies), discussed in Chapter 1. We begin with a generalization of the idea expressed in Figure 1.8.1.

For $1 \leq n \leq N$, let u^n denote the *unit vector in direction* n, which is defined as the vector in Z^N for which $u^n_n = 1$ and all other components are 0. A *basic digraph for* Z^N is a digraph (M, ρ) with the following properties:

(i) $M = \{\, u^n \mid 1 \leq n \leq N \,\} \cup \{\, -u^n \mid 1 \leq n \leq N \,\}$;
(ii) $\rho = \bigcup_{1 \leq i < j \leq N} \rho_{i,j}$, where the set of arcs $\rho_{i,j}$ is exactly one of the following:
  A. $\left\{ \left(u^i, u^j\right), \left(u^j, -u^i\right), \left(-u^i, -u^j\right), \left(-u^j, u^i\right) \right\}$,
  B. $\left\{ \left(u^i, -u^j\right), \left(-u^j, -u^i\right), \left(-u^i, u^j\right), \left(u^j, u^i\right) \right\}$,
  C. $\emptyset$.

Note that the definition of ρ permits selecting different options (A, B, or C) for specifying $\rho_{i,j}$ for different pairs of i and j.

As an example, consider Figures 1.7.4 and 1.8.1, which we now combine (indicating our new notation) into Figure 8.3.1. In this case $N = 3$, and so the number of nodes in the basic digraph is six. Also, there are three sets of arcs $\rho_{i,j}$ (since $1 \leq i < j \leq 3$), and here we have selected Option A for all three. Selecting Option B would correspond to reversing the direction of the corresponding cycle of arrows (e.g., choosing Option B for $\rho_{2,3}$ would correspond to reversing the horizontal cycle of arrows on the vertical faces of the sugar cube in Figure 8.3.1). As we have discussed in Chapter 1, such choices are arbitrary and make no

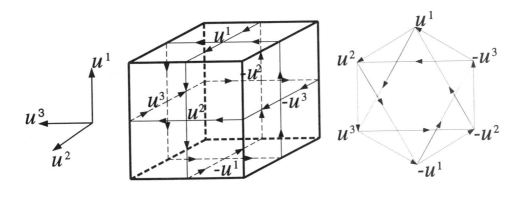

Figure 8.3.1. Illustration of a basic digraph and its relationship to the legitimate routes of the Fat Fly.

difference to the essential nature of what comes out of them. On the other hand, Option C is something new, it does not correspond to anything discussed in Chapter 1. The reason for including it will become clear later in this chapter.

The purpose of introducing basic digraphs is that they can be used to define adjacencies between bels in binary pictures over (Z^N, ω_N). Now we assume that we are given a basic digraph (M, ρ) and a binary picture (Z^N, ω_N, f) and explain how these are used to define an adjacency λ on the set of bels of (Z^N, ω_N, f). If $b = (c, d)$ is such a bel, then we have that $f(c) = 1$, $f(d) = 0$, and $(c, d) \in \omega_N$. From the definition of ω_N it follows that $d - c$ is a node of the given basic digraph (M, ρ). Every node ρ-adjacent from $d - c$ gives rise to a single bel λ-adjacent from b. Because of the utility of this in the proofs that follow, we define λ as $\bigcup_{1 \leq i < j \leq N} \lambda_{i,j}$, where every node $\rho_{i,j}$-adjacent from $d - c$ gives rise to a single bel $\lambda_{i,j}$-adjacent from b.

In following the details of the specification of how a node u which is $\rho_{i,j}$-adjacent from $d - c$ gives rise to a unique bel which is $\lambda_{i,j}$-adjacent from (c, d), consult Figure 8.3.2. (In that figure we are illustrating the situation when u has been selected using $\rho_{1,3}$. Three separate cases are considered: in (i) $d - c = -u^3$ and, hence, $u = u^1$, in (ii) $d - c = u^1$ and, hence, $u = u^3$, and in (iii) $d - c = u^3$ and, hence, $u = -u^1$.) We distinguish between three mutually exclusive possibilities. The first is that $d + u$ is a 1-spel (this corresponds to (iii) in Figure 8.3.2). In this case the bel $\lambda_{i,j}$-adjacent from (c, d) is specified to be $(d + u, d)$. The second is that $d + u$ is a 0-spel and $c + u$ is a 1-spel (this corresponds to (ii) in Figure 8.3.2). In this case the bel $\lambda_{i,j}$-adjacent from (c, d) is specified to be $(c + u, d + u)$. The third is that both $d + u$ and $c + u$ are 0-spels (this corresponds to (i) in Figure 8.3.2). In this case the bel $\lambda_{i,j}$-adjacent from (c, d) is specified to be $(c, c + u)$.

This definition provides a satisfactory reply to one of the questions raised in the last paragraph of the previous section: how easy is it to compute the bels λ-adjacent from a given

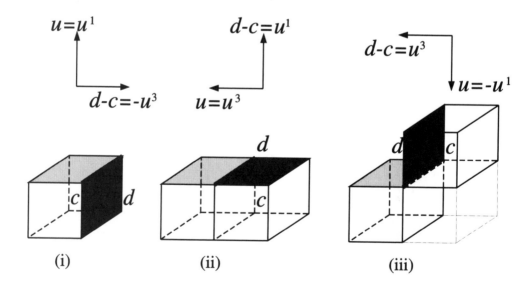

Figure 8.3.2. Illustration of the definition of bel-adjacency using the basic digraph of Figure 8.3.1. In each case, the darkly shaded bel is $\lambda_{1,3}$-adjacent to the lightly shaded one.

bel? For the definition given in this section the answer is, "it is very easy." A basic digraph is an easily represented object: for each of the $2N$ nodes, there are at most $N - 1$ nodes adjacent from it or to it. For any bel $b = (c, d)$, we have for some i $(1 \leq i \leq N)$ that $d - c \in \{u^i, -u^i\}$. For each (of the $N - 1$) $j \neq i$, there may (or may not) be a u that is $\rho_{i,j}$-adjacent from $d - c$. If there is such a u, it gives rise to a single bel $\lambda_{i,j}$-adjacent from (c, d), and the specification of this bel (as illustrated in Figure 8.3.2) requires checking the value of the binary function f at two spels at most.

The desirable behavior of the General Bel-Tracking Algorithm (as expressed by the theorem in the previous section) is guaranteed only if the adjacency λ is a bel-adjacency, i.e., λ^* is symmetric. Now we will show that this is the case for the $\lambda = \bigcup_{1 \leq i < j \leq N} \lambda_{i,j}$ as defined in this section, but first we prove a technical lemma of more general applicability.

Lemma 8.3.1. *Let ρ be a binary relation on a set M and L be a finite subset of M.*

  (i) *The transitive closure ρ^* of ρ is symmetric, provided that, for all $(c, d) \in \rho$, (d, c) is in ρ^*.*

  (ii) *If every element of L has exactly one element of L ρ-adjacent from it and at least one element of L ρ-adjacent to it, then every element of L has exactly one element of L ρ-adjacent to it.*

  (iii) *If every element of L has exactly one element of L ρ-adjacent from it and exactly one element of L ρ-adjacent to it, then for any c and d in L such that c is ρ-adjacent to d, d is ρ-connected in L to c.*

Proof. Assume that, for all $(c, d) \in \rho$, (d, c) is in ρ^*. We prove by induction on K that if $\langle c^{(0)}, \cdots, c^{(K)} \rangle$ is a ρ-path, then $(c^{(K)}, c^{(0)}) \in \rho^*$. This is clearly the case, by the definition of transitive closure, if $K = 0$. If $K > 0$, then $(c^{(K)}, c^{(K-1)}) \in \rho^*$ (by the assumption) and $(c^{(K-1)}, c^{(0)}) \in \rho^*$ (by the induction hypothesis). Since ρ^* is clearly transitive, $(c^{(K)}, c^{(0)}) \in \rho^*$. This proves (i).

Now we define the binary relation τ on L by $(c, d) \in \tau$ if, and only if, $c \in L$, $d \in L$, and $(c, d) \in \rho$. Since it is assumed in (ii) that every element of L has exactly one element of L ρ-adjacent from it, the number of elements in τ must be exactly the number of elements in L. This shows that there cannot be an element of L that has more than one element of L ρ-adjacent to it since otherwise we would have more elements in τ than in L (since it is also assumed that every element of L has at least one element of L ρ-adjacent to it).

To complete the proof, let us assume that the premise of (iii) is satisfied and c and d in L are such that c is ρ-adjacent to d. We define an infinite sequence of elements of L as follows: $c^{(0)} = d$ and, for $k \geq 1$, $c^{(k)}$ is the unique element of L that is ρ-adjacent from $c^{(k-1)}$. Since L is finite, there must be a smallest positive integer l such that $c^{(l)} = c^{(m)}$ for some $m < l$. If m were positive, then the uniqueness of the element of L that is ρ-adjacent to $c^{(l)} = c^{(m)}$ would imply that $c^{(l-1)} = c^{(m-1)}$, which would contradict the minimality of l. Hence, $c^{(l)} = c^{(0)} = d$. Again by the uniqueness of the element of L which is ρ-adjacent to $c^{(l)} = d$, we have that $c^{(l-1)} = c$, and so $\langle c^{(0)}, \cdots, c^{(l-1)} \rangle$ is a ρ-path in L from d to c. $\square$

Theorem 8.3.2. *Let $(M, \bigcup_{1 \leq i < j \leq N} \rho_{i,j})$ be a basic digraph for Z^N. Assume that the set of all bels B in the binary picture (Z^N, ω_N, f) is finite and not empty and, for $1 \leq i < j \leq N$, define the binary relation $\lambda_{i,j}$ on B as above. Let $B_{i,j}$ be the subset of those bels $b = (c, d)$ for which $d - c \in \{u^i, -u^i, u^j, -u^j\}$. Then, for $1 \leq i < j \leq N$ such that $\rho_{i,j} \neq \emptyset$ and for every bel b in $B_{i,j}$, the following hold.*

  (i) *There is one, and only one, bel b' that is $\lambda_{i,j}$-adjacent from b. Furthermore, $b' \in B_{i,j}$.*

(ii) *There is one, and only one, bel b' in $B_{i,j}$ that is $\lambda_{i,j}$-adjacent to b. Furthermore, b is $\lambda_{i,j}$-connected in $B_{i,j}$ to b'.*

Moreover, $\lambda = \bigcup_{1 \leq i < j \leq N} \lambda_{i,j}$ is a bel-adjacency for $\left(Z^N, \omega_N, f\right)$.

Proof. By the definition of $\rho_{i,j}$, there is one and only one u that is $\rho_{i,j}$-adjacent from $d - c$ and this u is in $\{u^i, -u^i, u^j, -u^j\}$. According to the definition of $\lambda_{i,j}$, u gives rise to a unique bel $b' = (c', d')$ that is $\lambda_{i,j}$-adjacent from (c, d). Looking at the three possibilities in the definition of $\lambda_{i,j}$, we see that $d' - c'$ is (respectively) $d - (d + u) = -u$, $(d + u) - (c + u) = d - c$, or $(c + u) - c = u$. In any case, $b' \in B_{i,j}$. This proves (i).

To prove (ii) first we distinguish between three mutually exclusive possibilities (corresponding to the three cases represented in Figure 8.3.3, with $b = (c, d)$ corresponding to the lightly shaded bel in that figure) and show that in each one there is at least one bel $b' = (c', d')$ in $B_{i,j}$ that is $\lambda_{i,j}$-adjacent to b. Again we denote by u the element of $\{u^i, -u^i, u^j, -u^j\}$ which is $\rho_{i,j}$-adjacent from $d - c$.

The first possibility is that $d - u$ is a 1-spel (this corresponds to (iii) in Figure 8.3.3). In this case we define $c' = d - u$ and $d' = d$. Then $b' = (c', d')$ is a bel, $d' - c' = u$, and so $b' \in B_{i,j}$. From the definition of $\rho_{i,j}$ it follows (since u is $\rho_{i,j}$-adjacent from $d - c$) that $u' = c - d$ is $\rho_{i,j}$-adjacent from $d' - c' = u$. Since $d' + u' = d + (c - d) = c$ is a 1-spel, the definition of $\lambda_{i,j}$ implies $b' = (c', d')$ is $\lambda_{i,j}$-adjacent to $(d' + u', d') = (c, d) = b$.

The second possibility is that $d - u$ is a 0-spel and $c - u$ is a 1-spel (this corresponds to (ii) in Figure 8.3.3). In this case we define $c' = c - u$ and $d' = d - u$. Then $b' = (c', d')$ is a bel, $d' - c' = d - c$, and so $b' \in B_{i,j}$. Since u is $\rho_{i,j}$-adjacent from $d - c = d' - c'$ and $d' + u = d$ is a 0-spel and $c' + u = c$ is a 1-spel, the definition of $\lambda_{i,j}$ implies $b' = (c', d')$ is $\lambda_{i,j}$-adjacent to $(c' + u, d' + u) = (c, d) = b$.

The third possibility is that both $d - u$ and $c - u$ are 0-spels (this corresponds to (i) in Figure 8.3.3). In this case we define $c' = c$ and $d' = c - u$. Then $b' = (c', d')$ is

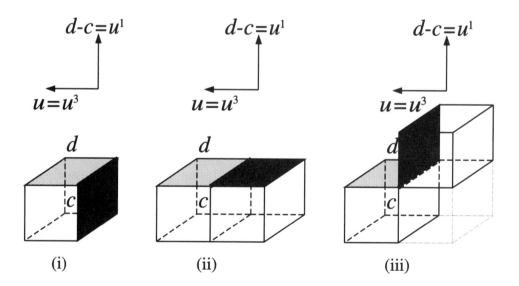

$$d\text{-}c = u^1 \qquad\qquad d\text{-}c = u^1 \qquad\qquad d\text{-}c = u^1$$

$$u = u^3 \qquad\qquad u = u^3 \qquad\qquad u = u^3$$

$$d \qquad\qquad\qquad d \qquad\qquad\qquad d$$

$$c \qquad\qquad\qquad c \qquad\qquad\qquad c$$

(i)   (ii)   (iii)

Figure 8.3.3. Illustration of Theorem 8.3.2(ii) using the basic digraph of Figure 8.3.1. In each case, the darkly shaded bel is $\lambda_{1,3}$-adjacent to the lightly shaded one.

a bel, $d' - c' = -u$, and so $b' \in B_{i,j}$. From the definition of $\rho_{i,j}$ it follows (since u is $\rho_{i,j}$-adjacent from $d - c$) that $u' = d - c$ is $\rho_{i,j}$-adjacent from $d' - c' = -u$. Since $d' + u' = (c - u) + (d - c) = d - u$ and $c' + u' = c + (d - c) = d$ are both 0-spels, the definition of $\lambda_{i,j}$ implies $b' = (c', d')$ is $\lambda_{i,j}$-adjacent to $(c', c' + u') = (c, d) = b$.

This shows that, for any bel b, there is at least one bel b' in $B_{i,j}$ that is $\lambda_{i,j}$-adjacent to b. According to (i), it is also the case that there is exactly one bel b' in $B_{i,j}$ that is $\lambda_{i,j}$-adjacent from b. By (ii) of the previous lemma, there is one and only one bel b' in $B_{i,j}$ that is $\lambda_{i,j}$-adjacent to b. Hence we can also apply (iii) of the previous lemma and find that b is $\lambda_{i,j}$-connected in $B_{i,j}$ to b'. This completes the proof of (ii).

To complete the proof, we use (i) of the previous lemma. If $(a, b) \in \lambda$, then $(a, b) \in \lambda_{i,j}$ for some $1 \leq i < j \leq N$ such that $\rho_{i,j} \neq \emptyset$. By (ii), this implies that b is $\lambda_{i,j}$-connected in $B_{i,j}$ to a and, consequently, $(b, a) \in \lambda^*$. Hence, λ^* is symmetric. $\square$

Corollary 8.3.3. *For $N \geq 2$, let (M, ρ) be a basic digraph for Z^N. Assume that the set of all bels B in the binary picture (Z^N, ω_N, f) is finite and define the binary relation λ on B as above. Then, for any bel i, the General Bel-Tracking Algorithm terminates in a finite number of steps and, at that time, S is that λ-component of the set of bels that contains i.*

The importance of this corollary is partially due to the fact that by using it we are able to complete the proofs of all outstanding claims regarding the behavior of the various boundary-tracking algorithms given in Chapter 1. Although we leave the details of this to the next section, we wish to complete the current section by discussing the relationship of the General Bel-Tracking Algorithm to Artzy's Algorithm. Essentially, we claim that the latter is a special case of the former.

The strict mathematical demonstration of such a claim is hampered by the fact that Artzy's Algorithm was introduced quite early in our book and its description is less formal than that of the General Bel-Tracking Algorithm. However, we can identify the informal language of Chapter 1 with the more rigorous discussion of the current chapter as follows.

Clearly, Artzy's Algorithm is applicable to binary pictures in the digital space (Z^3, ω_3). The 1-spels of this digital space are referred to in Chapter 1 as being occupied by sugar cubes. The faces separating voxels occupied by sugar cubes from voxels not occupied by sugar cubes are exactly the bels of the current chapter. If we define λ as above using the basic digraph of Figure 8.3.1 and Option A for all $1 \leq i \leq j \leq 3$, then we see (by comparing to Figure 1.10.1) that, for any bel a, the two bels to which the Cloning Fly would get from a are exactly the two bels λ-adjacent from a. Noting also that, for any bel b, $in_\lambda(b) = 2$, we see that if we make the following identification between the terminology of the General Bel-Tracking Algorithm (with λ as specified above) and Artzy's Algorithm, then the two are strictly identical:

(i) i is "the face meets the one on which you are standing at the edge in front of you";

(ii) L is the "queue";

(iii) to put something into S is to "dirty" it;

(iv) T is the "list".

In view of this, the previous corollary implies that under the conditions of Claim 1.11.2 Artzy's Algorithm terminates in a finite number of steps and at that time the set of dirty faces is exactly that λ-component of the set of bels which contains i.

8.4. Proofs of the Boundary-Tracking Claims

The characterization, given at the end of the previous section, of the set faces dirtied by the time Artzy's Algorithm stops is not the one given in Claim 1.11.2. To complete the proof of correctness of Claim 1.11.2, we need the following important result.

Theorem 8.4.1. *For $N \geq 2$, let $(M, \bigcup_{1 \leq i < j \leq N} \rho_{i,j})$ be a basic digraph for Z^N such that $\rho_{i,j} \neq \emptyset$ for any $1 \leq i < j \leq N$. Assume that the set of all bels B in the binary picture (Z^N, ω_N, f) is finite and define the binary relation λ on B as above. Then, for any bel $i = (c, d)$, the General Bel-Tracking Algorithm terminates in a finite number of steps and, at that time, $S = \partial(O, Q)$, where O is the δ_N-component of the set of 1-spels which contains c and Q is the ω_N-component of the set of 0-spels which contains d.*

Proof. By the previous corollary we know that the algorithm terminates in a finite number of steps and that at that time S is that λ-component of B which contains i.

First we show inductively that, both just before and just after the execution of Instruction (2) in the General Bel-Tracking Algorithm, $L \subset \partial(O, Q)$ and $S \subset \partial(O, Q)$. This is true just before to the first execution of Instruction (2), since only i is put into L and S due to Instruction (1) and, clearly, $i = (c, d) \in \partial(O, Q)$. Now assume the validity of the induction hypothesis just before some execution of Instruction (2). If a bel b is put into L and S during the execution of Instruction (2), then there must have been an a removed from $L \subset \partial(O, Q)$ such that b is $\lambda_{i,j}$-adjacent from a. By considering the three possible ways that a can be $\lambda_{i,j}$-adjacent to b (look at Figure 8.3.2), we see that $a \in \partial(O, Q)$ implies that $b \in \partial(O, Q)$. (Essential use is made here of the fact that O is a δ_N-component of the set of 1-spels.) Since nothing is put into L or S during the execution Instruction (3), this completes the induction.

Next we show that, at the termination of the General Bel-Tracking Algorithm, S is 2-locally-Jordan. By definition, we need to show that $p_S\langle c^{(0)}, c^{(1)}, c^{(2)}, c^{(3)} \rangle$ is odd for any ω_N-path such that $(c^{(0)}, c^{(3)}) \in S$. (Here we have used the fact that there is no ω_N-path of length 2 from a spel to a spel ω_N-adjacent from it.) The desired result

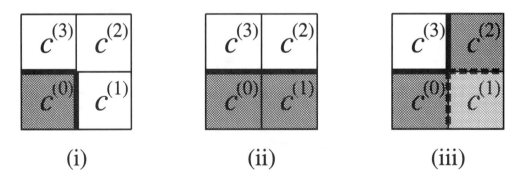

$\qquad\qquad$ (i) $\qquad\qquad\qquad\qquad\qquad$ (ii) $\qquad\qquad\qquad\qquad\qquad$ (iii)

Figure 8.4.1. Illustration of the three cases in the proof which shows that the output of the General Bel-Tracking Algorithm based on the basic digraph $(M, \bigcup_{1 \leq i < j \leq N} \rho_{i,j})$ with $\rho_{i,j} \neq \emptyset$, for any $1 \leq i < j \leq N$, is 2-locally-Jordan. The spels shaded dark gray are 1-spels. The spel shaded light gray may be either a 1-spel or a 0-spel. All other spels are 0-spels.

is trivially true if $c^{(0)} = c^{(2)}$ or $c^{(1)} = c^{(3)}$. When neither of these is the case, then the loop $\langle c^{(0)}, c^{(1)}, c^{(2)}, c^{(3)}, c^{(0)} \rangle$ is (by definition) a unit square and hence by Lemma 6.3.7 a unit lattice square. Looking at (6.3.6), we see that this implies that, for some $1 \le i < j \le N$, $c^{(3)} - c^{(0)} = c^{(2)} - c^{(1)} \in \{u^i, -u^i, u^j, -u^j\}$ and $c^{(3)} - c^{(2)} = c^{(0)} - c^{(1)} \in \{u^i, -u^i, u^j, -u^j\}$; see Figure 8.4.1. In that figure we distinguish among three cases, corresponding to the three cases of Figure 8.3.2. In each case, the bel $(c^{(0)}, c^{(3)})$ in Figure 8.4.1 corresponds to the bel painted light gray in Figure 8.3.2. In Case (i), $f(c^{(1)}) = f(c^{(2)}) = 0$. Both the bels $(c^{(0)}, c^{(3)})$ and $(c^{(0)}, c^{(1)})$ are in $B_{i,j}$ (as defined in the statement of the previous theorem) and, since $\rho_{i,j} \ne \emptyset$, one of them has to be $\lambda_{i,j}$-adjacent to the other. The previous theorem implies that they are in the same λ-component of B. Since $(c^{(0)}, c^{(3)}) \in S$, this implies that $(c^{(0)}, c^{(1)}) \in S$. On the other hand, $(c^{(1)}, c^{(2)})$ and $(c^{(2)}, c^{(3)})$ are not bels and so cannot be in S. This proves that $p_S \langle c^{(0)}, c^{(1)}, c^{(2)}, c^{(3)} \rangle = 1$ in Case (i). The proof in Case (ii) is very similar and therefore is not given. In Case (iii) there is an additional complication. One can argue, as in Case (i), to show that in Case (iii) $(c^{(2)}, c^{(3)}) \in S$, but the possibilities of $(c^{(0)}, c^{(1)}) \in S$ or $(c^{(1)}, c^{(2)}) \in S$ need also be investigated. Neither of these is true if $c^{(1)}$ is a 1-spel. On the other hand, if $c^{(1)}$ is a 0-spel, then both $(c^{(0)}, c^{(1)})$ and $(c^{(1)}, c^{(2)})$ are in $B_{i,j}$ and, consequently, either both are in S or neither is in S. In any case, again $p_S \langle c^{(0)}, c^{(1)}, c^{(2)}, c^{(3)} \rangle$ is odd.

Next we show that, at the termination of the General Bel-Tracking Algorithm, S is near-Jordan. We use Lemma 6.2.4. We have just shown that S is a 2-locally-Jordan surface in the digital space (Z^N, ω_N), which is 2-simply connected by Theorem 6.3.5. Since S is a set of bels, it has to be antisymmetric. To show that it is near-Jordan it is sufficient to show (according to Lemma 6.2.4) that, for any two elements in the immediate exterior of S, there is an ω_N-path between them which does not cross S. Suppose that g and h are in $IE(S)$. That implies that there exist spels g' and h' such that both (g', g) and (h', h) are in S. Since S is a λ-component of B, there is λ-path in S from (g', g) to (h', h). Since a λ-path is a sequence of consecutively λ-adjacent bels, the task of this paragraph is done if we can show that, whenever a bel (g', g) is λ-adjacent to a bel (h', h), then there is an ω_N-path from g to h that does not cross B. This follows since (g', g) must be $\lambda_{i,j}$-adjacent to (h', h), for some $1 \le i < j \le N$, and in each of the three cases in the definition of $\lambda_{i,j}$ the required result follows trivially. (For example, the three relevant ω_3-paths in Figure 8.3.2 are (i) $\langle d, d + u, c + u \rangle$, (ii) $\langle d, d + u \rangle$, and (iii) $\langle d \rangle$.)

Now we complete the proof by showing that, at the termination of the General Bel-Tracking Algorithm, $S = \partial(O, Q)$. We already know that $S \subset \partial(O, Q)$. To prove the converse, consider any $(g, h) \in \partial(O, Q)$, i.e., $g \in O$, $h \in Q$, $(g, h) \in \omega_N$. Since $g \in O$, there is a δ_N-path in O from c to g from which we can create an ω_N-path $\langle c^{(0)}, \cdots, c^{(K)} \rangle$ from c to g with the following property: for $0 \le k \le K$, either $f(c^{(k)}) = 1$ or $f(c^{(k)}) = 0$ and $f(c^{(k-1)}) = f(c^{(k+1)}) = 1$. In the latter case, $(c^{(k-1)}, c^{(k)})$ and $(c^{(k+1)}, c^{(k)})$ are bels such that one of them is λ-adjacent to the other, and so either both are in S or neither is in S. Therefore it follows that $p_S \langle c^{(0)}, \cdots, c^{(K)} \rangle = 0$. Since $h \in Q$, there is an ω_N-path $\langle d^{(0)}, \cdots, d^{(L)} \rangle$ in Q from h to d. Clearly $p_S \langle d^{(0)}, \cdots, d^{(L)} \rangle = 0$. It follows that $(g, h) \in S$, for otherwise the crossing parity through the near-Jordan surface S would be even for the ω_N-path $\langle c = c^{(0)}, \cdots, c^{(K)} = g, h = d^{(0)}, \cdots, d^{(L)} = d \rangle$, contradicting Theorem 6.2.1. $\square$

Corollary 8.4.2. *Claim 1.11.2 is valid.*

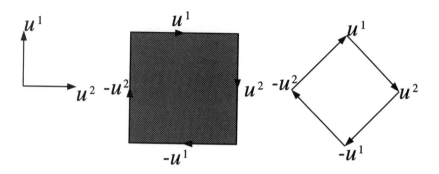

Figure 8.4.2. The basic digraph for Z^2 using Option A for $\rho_{1,2}$ is shown on the right. For the coordinate system in which u^1 and u^2 are as indicated on the left, this basic digraph corresponds to the "legitimate route" on a flat sugar cube, as indicated in the middle.

Although the validity of Artzy's Algorithm, as expressed in Claim 1.11.2, is an important result by itself, we note that the theorem of which it is a corollary is of a more general applicability: it proves the validity of the generalization of Artzy's Algorithm to $\left(Z^N, \omega_N\right)$ with an arbitrary $N \geq 2$. In fact, the same theorem can also be used to prove the validity of other algorithms of Chapter 1, as we now demonstrate.

If (M, ρ) is a basic digraph for Z^2, then $M = \{(1,0), (0,1), (-1,0), (0,-1)\}$. If we choose Option A for $\rho_{1,2}$, then

$$\rho = \{((1,0), (0,1)), ((0,1), (-1,0)), ((-1,0), (0,-1)), ((0,-1), (1,0))\}. \qquad (8.4.1)$$

This basic digraph is illustrated in Figure 8.4.2. Now we illustrate that with the bel-adjacency $\lambda = \lambda_{1,2}$, defined as above based on this ρ, the General Bel-Tracking Algorithm is essentially the same as the Algorithm for Flat Flies. (This is for the Flat Fly placed on a flat sugar cube facing in the direction indicated in Figure 1.3.1. If the Flat Fly is placed facing in the opposite direction, then we would have to use Option B for $\rho_{1,2}$ to obtain the corresponding result.)

We note that, for any bel b of any binary picture (Z^2, ω_2, f), there is exactly one bel λ-adjacent to it, i.e., $in_\lambda(b) = 1$. Also, there is exactly one bel λ-adjacent from b. We use the following fact: if the fly is standing on a bel $b = (c, d)$, then the bel onto which it crawls due to Instruction (2) of the Algorithm for Flat Flies is exactly the unique bel λ-adjacent from b. This is demonstrated in Figure 8.4.3 for the first three steps of the illustration of the Algorithm for Flat Flies in Figure 1.3.2. On the left of Figure 8.4.3, $d + u$ is a 0-spel and $c + u$ is a 1-spel, therefore (c, d) is λ-adjacent to $(c + u, d + u)$. In the middle, $d + u$ is a 1-spel, therefore (c, d) is λ-adjacent to $(d + u, d)$. On the right, both $d + u$ and $c + u$ are 0-spels, therefore (c, d) is λ-adjacent to $(c, c + u)$.

In view of this, it is easy to give an inductive proof of the following relationship between the behavior of the General Bel-Tracking Algorithm with the λ defined using the basic digraph of Figure 8.4.2 and i specified as the initial location of the Flat Fly and the behavior or the Algorithm for Flat Flies. The number of times K that Instruction (2) is executed by the two algorithms is the same and just before the kth execution of Instruction (2) in the respective algorithms ($1 \leq k \leq K$) the following are true.

(i) S is the set of dirty faces.

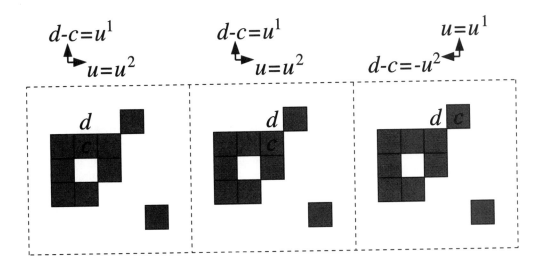

Figure 8.4.3. The behavior of the General Bel-Tracking Algorithm with the λ defined using the basic digraph of Figure 8.4.2. This illustration corresponds to the top row of Figure 1.3.2.

(ii) L is the singleton set containing the bel on which the Flat Fly is standing.

(iii) $T = \{i\}$.

Just before the Kth execution of Instruction (2) in the respective algorithms, the Flat Fly will be standing on the unique bel λ-adjacent to i. Therefore both Algorithms will stop in the next execution of Instruction (3) without changing S or dirtying any faces, respectively. Hence, the set of faces dirtied by the Algorithm for Flat Flies is exactly S at the termination of the General Bel-Tracking Algorithm. Thus our previous theorem implies the following corollary, which states that the Algorithm for Flat Flies performs as claimed.

Corollary 8.4.3.   *Claim 1.5.1 is valid.*

To prove the rest of the claims in Chapter 1, we need to consider the Inefficient Bel-Tracking Algorithm. This is because if we repeat the argument given in the final paragraph of the preceding section, but with the Algorithm Simulating Cloning Flies replacing Artzy's Algorithm and the Inefficient Bel-Tracking Algorithm replacing the General Bel-Tracking Algorithm, then we arrive at the conclusion that the Inefficient Bel-Tracking Algorithm (with the specified λ) and the Algorithm Simulating Cloning Flies are strictly identical. This, together with the previously established identification of the General Bel-Tracking Algorithm (with the specified λ) with Artzy's Algorithm, leads to the following consequence of Theorem 8.2.4.

Corollary 8.4.4.   *Claim 1.11.1 is valid.*

Now if we combine the already validated Claims 1.11.1 and 1.11.2 and also consider Theorem 1.10.6, then we see that the following is the case.

Corollary 8.4.5.   *Claims 1.10.3 and 1.10.1 are valid.*

Now we have taken care of all the previously unproven claims of the earlier chapters. Prior to going on to the discussion of efficient boundary tracking in the fcc grid, we mention three more topics to do with boundary tracking in (Z^N, ω_N).

Our intuitive model, in which we represent 1-spels by sugar cubes, channeled us in the direction of tracking $\delta_N \omega_N$-boundaries. From a purely mathematical point of view, $\omega_N \delta_N$-boundaries have the same validity. It turns out that the theory presented in this and the previous two sections is easily adapted to tracking $\omega_N \delta_N$-boundaries. (Indeed, if this were not so, then we would have considered the whole approach of this book a failure since our aim was to develop a theory easily applicable to a wide variety of problems in the geometry of digital spaces.) In fact, the only change that needs to be made in the previous development is in the specification of how a node u which is $\rho_{i,j}$-adjacent from $d - c$ gives rise to a unique bel that is $\lambda_{i,j}$-adjacent from (c, d). Again we distinguish among three mutually exclusive possibilities. The first is that $c + u$ is a 0-spel (this corresponds to (i) in Figure 8.3.2). In this case the bel $\lambda_{i,j}$-adjacent from (c, d) is specified to be $(c, c + u)$. The second is that $c + u$ is a 1-spel and $d + u$ is a 0-spel (this corresponds to (ii) in Figure 8.3.2). In this case the bel $\lambda_{i,j}$-adjacent from (c, d) is specified to be $(c + u, d + u)$. The third is that both $c + u$ and $d + u$ are 1-spels (this corresponds to (iii) in Figure 8.3.2). In this case the bel $\lambda_{i,j}$-adjacent from (c, d) is specified to be $(d + u, d)$. We leave it to the reader to state and prove the analog of Theorem 8.4.1 for this bel-adjacency and $\omega_N \delta_N$-boundaries.

The second topic has to do with choosing Option C in the definition of the basic digraph. For example, we may pick an arbitrary n, $1 \leq n \leq N$, and choose $\rho_{i,j} = \emptyset$ if, and only if, $n \notin \{i, j\}$. This is illustrated in Figure 8.4.4 for the case $N = 3$ and $n = 1$. Why would we want to do such a thing? The answer is that such a choice of the basic digraph results in a version of the General Bel-Tracking Algorithm which is even more efficient than Artzy's Algorithm. This comes about as follows. For any bel $b = (c, d)$ for which $d - c \notin \{u^n, -u^n\}$, there will be one, and only one, bel that is λ-adjacent from b. If, while executing Instruction (2)b of the General Bel-Tracking Algorithm, such a bel b is put into L and S, no copies of it are put into T (since $in_\lambda(b) = 1$). Even without any further examination, we see that this is likely to lead to increased efficiency since the size of T is likely to be much smaller than in the previously discussed algorithms. (Assuming an even distribution of bels in the six possible different orientations in three-dimensional euclidean space, we see from Figure 8.4.4 that only one third of the bels that at one time or other are put into T using Artzy's Algorithm are ever put into T using the new basic digraph. The efficiency increase is even greater in higher dimensional spaces.) However, we can achieve a further increase in efficiency by paying some special attention to the queuing discipline for L. Since the proven validity of the General Bel-Tracking Algorithm is independent of the queuing discipline for L, further saving in computational cost may result from putting bels of the type that we are discussing above only in S and immediately moving to the beginning of Instruction (2) just as if this bel were just removed from L. We forego a detailed discussion of the behavior of this algorithm but point out that in addition to its termination at a time when S is a λ-component of the set of bels (as guaranteed by the final corollary of the previous section), we also know [22, 26] that this final S is a $\gamma_n \omega_N$-boundary (if λ is defined from ρ, as we have done earlier) or an $\omega_N \gamma_n$-boundary (if λ is defined from ρ as in the previous paragraph). Here γ_n is the spel-adjacency in (Z^N, ω_N) defined on page 147. As stated at the same place, $\{\gamma_n, \omega_N\}$ is a Jordan pair, and so the output S of the efficient boundary tracking algorithm discussed in this

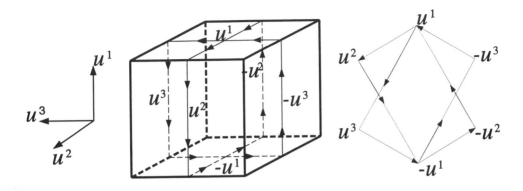

Figure 8.4.4. Illustration of a basic digraph with $N = 3$ and $\rho_{2,3} = \emptyset$, together with the corresponding routes of the Fat Fly.

paragraph is guaranteed to be a Jordan surface. The only negative aspect of this algorithm is its anisotropic nature: depending on the choice of n, different surfaces may be detected starting with the same bel in the same binary picture.

Finally, we point out that the General Bel-Tracking Algorithm can be adapted to track $\alpha_N\omega_N$-boundaries in (Z^N, ω_N). We briefly discuss the idea of an efficient implementation by looking at an adaptation of Artzy's Algorithm to tracking $\alpha_3\omega_3$-boundaries. The simple trick is to use a queuing discipline so that, given a bel i, we first track the $\delta_3\omega_3$-boundary that contains i. When this has been done, the queue may contain some bels which are in the $\alpha_3\omega_3$-boundary that contains i but not in the $\delta_3\omega_3$-boundary that contains i. Picking one of these, we track the $\delta_3\omega_3$-boundary that contains it, and so on. When the queue is empty, the $\alpha_3\omega_3$-boundary has been detected, and the essential efficiency of Artzy's Algorithm has been retained.

8.5. Boundary Tracking in the FCC Grid

We recall that (F, β_1) is a finitary 1-simply connected digital space. (Finitariness is obvious and, in view of Exercise 6.6, the rest follows from Theorems 3.4.3 and 6.4.5.) That means that we immediately get the following from the results of the first section of this chapter.

Corollary 8.5.1. *Given any binary picture (F, β_1, f) and a 1-spel c and a 0-spel d such that $(c, d) \in \beta_1$ and the boundary $\partial(O, Q)$ between the β_1-component O of 1-spels containing c and the β_1-component Q of 0-spels containing d is finite, the 1-Simply Connected Algorithm terminates in a finite number of steps and at that time $S = \partial(O, Q)$.*

At first glance this may imply that no more needs to be said regarding boundary detection in the fcc grid, especially since we also know that the output S mentioned in the corollary is a Jordan surface. However, the efficiency of the 1-Simply Connected Algorithm is questionable since its general nature is similar to the Inefficient Bel-Tracking Algorithm rather than to the General Bel-Tracking Algorithm. What we show in this section is that a simple modification of Artzy's Algorithm suffices to produce an algorithm for efficient boundary tracking in both the cubic and the fcc grid.

First we note the following relationship between the cubic grid Z^3 and the fcc grid F.

Lemma 8.5.2.   *For any c and d in F,*

$$(c,d) \in \beta_1 \Leftrightarrow (c,d) \in \delta_3 . \tag{8.5.1}$$

Proof. Assume that $(c,d) \in \beta_1$. This means (see page 46) that c and d differ by 1 in two of the coordinates and are identical in the third, which implies that $(c,d) \in \delta_3$.

On the other hand, if both c and d are in F, then it cannot possibly be that $(c,d) \in \omega_3$. Hence $(c,d) \in \delta_3$ must imply that c and d differ by 1 in two of the coordinates and are identical in the third, and so $(c,d) \in \beta_1$. $\square$

Now we come to an extremely interesting observation, which was made to this author by Hava Katz. The essential idea comes from the superimposition, shown in Figure 8.5.1, of the right-hand sides of Figures 1.2.1 and 2.1.2. We notice that, for those elements in the cubic grid Z^3 which also belong to the fcc grid F, there is a one-to-one correspondence between the edges of the cubic Voronoi neighborhoods of cubic grid and the faces of the rhombic dodecahedral Voronoi neighborhoods of the fcc grid. This, combined with the observation (see Figure 1.7.2) that an algorithm can achieve only what is required in Aim 1.6.1 if the Fat Fly traverses each edge of a single sugar cube, leads us to the following mathematical development.

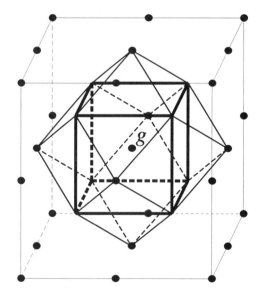

Figure 8.5.1. Superimposition of the right-hand sides of Figures 1.2.1 and 2.1.2.

First we formally introduce the face-or-edge-or-vertex-adjacency for the fcc grids that was already briefly mentioned in Chapter 2. This is the spel-adjacency in (F, β_1) which is defined as

$$\overline{\beta_1} = \{ (c, d) \mid 0 < \|c - d\| \leq 2 \} . \tag{8.5.2}$$

Next, for any binary picture (F, β_1, f) over the fcc grid, we define the *derived binary picture* (Z^3, ω_3, f') over the cubic grid by

$$f'(c) = \begin{array}{l} f(c), \text{ if } c \in F, \\ 0, \quad \text{otherwise.} \end{array} \tag{8.5.3}$$

Now these two new concepts are used in the following preliminary result.

Lemma 8.5.3. *Let* (Z^3, ω_3, f') *be the binary picture derived from* (F, β_1, f).

(i) For any 1-spel c of (F, β_1, f), let O be the β_1-component that contains c of the set of 1-spels of (F, β_1, f) and let O' be the δ_3-component that contains c of the set of 1-spels of (Z^3, ω_3, f'). Then $O = O'$.

(ii) For any 0-spel d of (F, β_1, f), let Q be the $\overline{\beta_1}$-component that contains d of the set of 0-spels of (F, β_1, f) and let Q' be the ω_3-component that contains d of the set of 0-spels of (Z^3, ω_3, f'). Then $Q = Q' \cap F$.

Proof. It follows immediately from the previous lemma that any subset of the set of 1-spels of (F, β_1, f) is β_1-connected if, and only if, it is δ_3-connected. Therefore it follows that the set of 1-spels of (F, β_1, f), which is the same as the set of 1-spels of (Z^3, ω_3, f'), has the same partitioning into β_1-components as into δ_3-components; which proves (i).

To prove (ii), we make the observation (whose easy proof we leave to the reader) that for any two distinct elements u and w of the fcc grid F, $(u, w) \in \overline{\beta_1}$ if, and only if, there is an element v of $Z_3 - F$ that is ω_3-adjacent to both u and w.

First suppose that $e \in Q$. Since Q is a $\overline{\beta_1}$-component of the set of 0-spels of (F, β_1, f) containing d, there exists a $\overline{\beta_1}$-path $\langle c^{(0)}, \cdots, c^{(K)} \rangle$ of 0-spels of (F, β_1, f) from d to e. To show that $e = c^{(K)} \in Q' \cap F$, we show by induction that $c^{(k)} \in Q' \cap F$ for $0 \leq k \leq K$. This is clearly the case for $c^{(0)} = d$. Now assume that $1 \leq k \leq K$ and $c^{(k-1)} \in Q' \cap F$. Since $c^{(k-1)}$ is $\overline{\beta_1}$-adjacent to $c^{(k)}$, it follows from the observation above that there is an element v of $Z_3 - F$ that is ω_3-adjacent to both $c^{(k-1)}$ and $c^{(k)}$. Since v is not in F, $f'(v) = 0$. Hence $c^{(k-1)}$, v, and $c^{(k)}$ are in the same ω_3-component (namely in Q') of the 0-spels (Z^3, ω_3, f'). Hence $c^{(k)} \in Q' \cap F$.

Conversely, suppose that $e \in Q' \cap F$. Since Q' is an ω_3-component of the set of 0-spels of (Z^3, ω_3, f') containing d, there exists a ω_3-path $\langle c^{(0)}, \cdots, c^{(K)} \rangle$ in the set of 0-spels of (Z^3, ω_3, f') from d to e. By the definitions of ω_3 and of $F = F_1$, it is clear that $c^{(k)} \in F$ if, and only if, k is even. (In particular, this implies that K must be even.) To complete the proof, we show by induction that $c^{(k)} \in Q$ for every even k such that $0 \leq k \leq K$. This is clearly the case for $c^{(0)} = d$. Now assume that $2 \leq k \leq K$ and $c^{(k-2)} \in Q$. Since $(c^{(k-2)}, c^{(k-1)}) \in \omega_3$ and $(c^{(k-1)}, c^{(k)}) \in \omega_3$, it follows from the observation above that either $c^{(k)} = c^{(k-2)}$ or $(c^{(k-2)}, c^{(k)}) \in \overline{\beta_1}$. In either case, $c^{(k)} \in Q$. $\square$

This leads us to our very last topic, the presentation of a variant of Artzy's Algorithm capable of boundary tracking in both the cubic and the fcc grid and the proof of its validity.

The assumption is that in case the boundary tracking is to be done in a binary picture (F, β_1, f) over the fcc grid, then first we transform this into the derived binary picture (Z^3, ω_3, f') over the cubic grid. In either case, the initially specified bel (c^0, d^0) is a bel in the (original or derived) binary picture over the cubic grid, and the bel-adjacency λ is the one that makes the General Bel-Tracking Algorithm strictly identical to Artzy's Algorithm.

Tracking Algorithm for Cubic and FCC Grids

(1) Put (c^0, d^0) into L and S_{Z^3} and put two copies of (c^0, d^0) into T.

(2) Remove a bel (c, d) from L. Let (c^1, d^1) and (c^2, d^2) be the two bels that are λ-adjacent from (c, d). For $k \in \{1, 2\}$, do the following. If $c^k = c$, then put $(c, d + d^k - c)$ into S_F. Try to find one copy of (c^k, d^k) in T.

    a.   If successful, remove this copy of (c^k, d^k) from T.

    b.   If not, then put (c^k, d^k) into L, S_{Z^3}, and T.

(3) Check if L is empty.

    a.   If it is, STOP.

    b.   If it is not, start again at Instruction (2).

Theorem 8.5.4.   *Let (Z^3, ω_3, f') be a binary picture with finitely many bels, and let (c^0, d^0) be one such bel.*

(i) *The Tracking Algorithm for Cubic and FCC Grids terminates in a finite number of steps, and at that time $S_{Z^3} = \partial_{\omega_3}(O', Q')$, where O' is the δ_3-component that contains c^0 of the set of 1-spels of (Z^3, ω_3, f') and Q' is the ω_3-component that contains d^0 of the set of 0-spels of (Z^3, ω_3, f').*

(ii) *If (Z^3, ω_3, f') is the binary picture derived from (F, β_1, f) and there is a d' in Z^3 such that (c^0, d') is a bel of (Z^3, ω_3, f') that is λ-adjacent from (c^0, d^0), then $(c^0, d^0 + d' - c^0)$ is a bel of (F, β_1, f), and at the time the Tracking Algorithm for Cubic and FCC Grids terminates, $S_F = \partial_{\beta_1}(O, Q)$, where O is the β_1-component that contains c^0 of the set of 1-spels of (F, β_1, f) and Q is the $\overline{\beta_1}$-component that contains $d^0 + d' - c^0$ of the set of 0-spels of (F, β_1, f).*

Proof. (i) is an immediate consequence of Theorem 8.4.1. This is because, if we ignore the output S_F of the Tracking Algorithm for Cubic and FCC Grids, then we see that it is identical to the General Bel-Tracking Algorithm (when we identify the i and S of that algorithm with the (c^0, d^0) and S_{Z^3}, respectively, of the Tracking Algorithm for Cubic and FCC Grids).

Now assume that (Z^3, ω_3, f') is the binary picture derived from (F, β_1, f) and there is a d' in Z^3 such that (c^0, d') is a bel of (Z^3, ω_3, f') that is λ-adjacent from (c^0, d^0). Looking at the definition of λ, we see that the only way (c^0, d') can be λ-adjacent from (c^0, d^0) is if the applicable case is the one illustrated in (i) of Figure 8.3.2. This case can arise only if $f'(d^0 + d' - c^0) = 0$. Since $f'(c^0) = 1$, we know that $c^0 \in F$ and, consequently, $d^0 + d' - c^0 \in F$. It follows that $(c^0, d^0 + d' - c^0)$ is a bel of (F, β_1, f). It follows immediately from (i) of the previous lemma that $O = O'$. Also, since d^0 and $d^0 + d' - c^0$ are ω_3-adjacent 0-spels of (Z^3, ω_3, f'), it follows from (ii) of the previous lemma that $Q = Q' \cap F$.

A bel that in S_F at the termination of the Tracking Algorithm for Cubic and FCC Grids must have been put into S_F while executing Instruction (2) and therefore must be of the form $(c, d + d^k - c)$, where the bel (c, d) of (Z^3, ω_3, f') (which has just been removed from L) is λ-adjacent to the bel (c, d^k) of (Z^3, ω_3, f'). This can happen only in the case illustrated in (i) of Figure 8.3.2 which can arise only if $f'(d + d^k - c) = 0$. Also, since (c, d) has been previously put into L, at the same time it must have been put into S_{Z^3}. By the already proved part (i) of the current theorem, therefore, we have that $c \in O' = O$ and $d \in Q'$, the latter of which implies (since $d + d^k - c$ is in F and is ω_3-adjacent to d) that $d + d^k - c$ is in $F \cap Q' = Q$. Thus we have just shown that $S_F \subset \partial_{\beta_1}(O, Q)$.

Conversely, suppose that $(c, d) \in \partial_{\beta_1}(O, Q)$. This implies that $c \in O'$, $d \in Q'$, and, for some $1 \leq i < j \leq 3$, $|d_i - c_i| = |d_j - c_j| = 1$ and $|d_n - c_n| = 0$ if $n \notin \{i, j\}$. Now we define two spels d^k ($k \in \{i, j\}$), both of which are ω_3-adjacent to both c and d, by $d^k_k = d_k$ and $d^k_n = c_n$ if $n \neq k$. Since d^k ($k \in \{i, j\}$) is not in F, $f'(d^k) = 0$ and, consequently, $d^k \in Q'$, and so $(c, d^k) \in \partial_{Z_3}(O', Q')$. We also observe that $f'(d^i + d^j - c) = f'(d) = f(d) = 0$, from which it follows that one of the bels (c, d^i) and (c, d^j) is $\lambda_{i,j}$-adjacent to the other. By part (i) of the current theorem, at some point that bel will be put into S_{Z_3} and hence into L by the Tracking Algorithm for Cubic and FCC Grids. When it is eventually removed from L, it will be discovered that the other one is λ-adjacent from it and at that time $(c, d^i + d^j - c) = (c, d)$ will be put into S_F. □

The behavior of the Tracking Algorithm for Cubic and FCC Grids has an interesting interpretation using the intuitive notions of sugar cubes and Fat Flies when the algorithm is applied to a finite picture in the fcc grid. In the fcc grid, the sugar cubes which occupy the Voronoi neighborhoods of 1-spels have the shape of rhombic dodecahedra. The relationship of these sugar cubes to the cube-shaped sugar cubes of the derived binary picture is illustrated in Figure 8.5.1. Note that each face of the cubic sugar cube corresponds to a vertex of the rhombic dodecahedral sugar cube. In a derived binary picture, the only type of λ-adjacencies that may exist correspond to Cases (i) and (iii) in Figure 8.3.2. (Case (ii) cannot happen, since two 1-spels of a derived binary picture cannot be ω_3-adjacent.) In Case (i), the Fat Fly of the derived binary picture crawls from one cubic face onto another, traversing (the heavily drawn) edge. This can also be interpreted as the Fat Fly of the fcc grid crawling along the diagonal of the rhombic face (which corresponds to the heavily drawn cubic edge) from one vertex to another. On the other hand, Case (iii) of Figure 8.3.2 does not produce an element of S_F since in such a situation there is no fcc bel corresponding to the heavily drawn cubic edge. Another way of saying this is that the Fat Fly of the fcc grid stays put at the common vertex corresponding to the two faces traversed by the Fat Fly of the cubic grid. So although an algorithm for Fat Flies for the cubic grid had the inherent inefficiency that to visit each bel at least once the fly had to visit each of them twice (recall Figure 1.7.2), this is not the case in the fcc grid: in the Fat Fly interpretation of the Tracking Algorithm for Cubic and FCC Grids, when the algorithm is applied to the fcc grid, each tracked bel is traversed exactly once.

We illustrate the output of the Tracking Algorithm for Cubic and FCC Grids on CT data of a patient's head. The original CT slices were real-valued pictures on the square grid $(0.8 \text{ mm})Z^2$, but the spacing between the slices in the third direction was 3.0 mm. (This is quite a typical mode of data collection in CT.) To produce real-valued pictures over a cubic grid and over an fcc grid, we have estimated (by interpolation) what the values would be at the points of such grids. We have chosen the cubic grid to be $(0.65 \text{ mm})Z^3$. On the basis that one reasonable way of comparing the outputs of the Tracking Algorithm for Cubic and FCC Grids on the two different kinds of grid is to make the size of the faces of the Voronoi neighborhoods the same, we have selected the fcc grid to be F_ϕ with $\phi = 2^{0.25} \times 0.65$ mm; see Exercise 8.16. As can be seen from the same exercise, in this case the volumes of the rhombic dodecahedra are more than 3.36 times larger than the volumes of the cubes. Consequently, the number of fcc grid points needed to cover a particular volume of space is

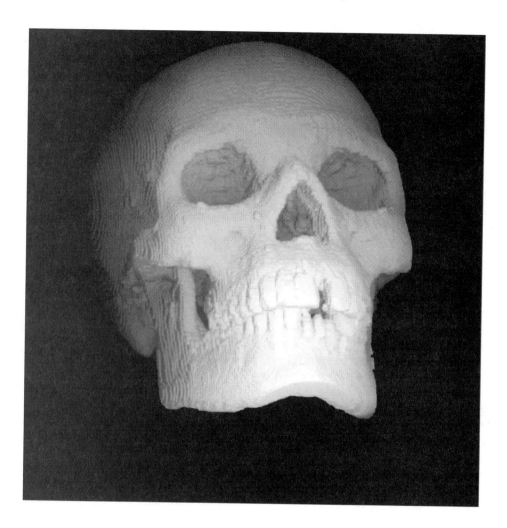

Figure 8.5.2. Three-dimensional display of the skull of a patient approximated as a collection of 1,620,365 square-shaped faces of cubic voxels. The area of the individual faces is the same as in Figure 8.5.3.

less than a third of the number of cubic grids needed for the same purpose. Since bone in CT can be identified by thresholding, we used this method (with the same threshold for the two real-valued images in their respective digital spaces) to assign 1 to those grid points which correspond to bone and 0 to all other grid points. Then we applied the Tracking Algorithm for Cubic and FCC Grids to detect the digital approximations (in the two spaces) of the skull of the patient. The resulting surfaces are displayed in Figures 8.5.2 and 8.5.3. (We took a bit of liberty here. The Tracking Algorithm for Cubic and FCC Grids has been specified for the cubic grid Z^3 and for the fcc grid F_1, and yet here we are applying it to another cubic grid and to another fcc grid. That this is legitimate is a consequence of the isomorphism between any two cubic grids and any two fcc grids.)

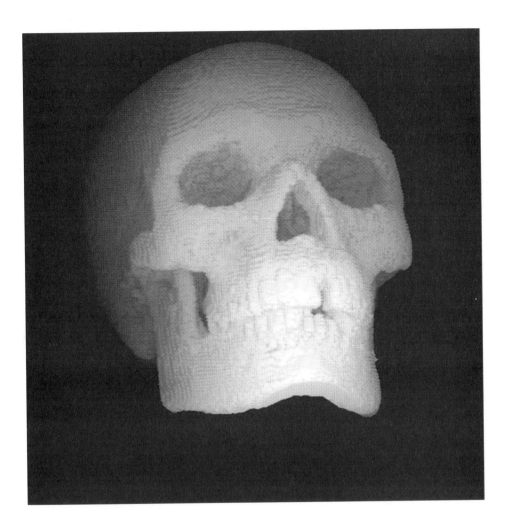

Figure 8.5.3. Three-dimensional display of the skull of a patient approximated as a collection of 1,213,125 rhombus-shaped faces of rhombic dodecahedral voxels. The area of the individual faces is the same as in Figure 8.5.2.

The first thing to note about these figures is that (even though the same set of CT slices, the same threshold, and the same area for the faces were used to produce the surfaces) the number of faces in the digital approximation to the skull based on the cubic grid is approximately 25% more than the number faces in the digital approximation based on the fcc grid. This is because the larger number of orientations of the rhombic dodecahedral faces allows us to fit the underlying biological surface more tightly. This also has a consequence on computational costs: the time required by Tracking Algorithm for Cubic and FCC Grids to detect the surface displayed in Figure 8.5.2 is nearly twice as long as the time required by the same algorithm to detect the surface displayed in Figure 8.5.3.

In displaying the surfaces we have chosen a rendering methodology [5] which does not hide the digital nature of the surfaces in question. We have done this so that the reader can observe the appearance of the digital surfaces themselves. From the same data one can produce much smoother appearing surfaces (see, for example, [43]), which, presumably, are more realistic renderings of the underlying biological surfaces. The question as to which surface (or, for that matter, which rendering method) is "better" should be answered with another question: "better for what?". Since, presumably, images such as those shown in Figures 8.5.2 and 8.5.3 are produced to help with a medical task, their "betterness" should be judged from the point of view of their efficacy for that task. Since a digital approximation is not perfect, the essential question is whether or not its imperfections hide the information that is important for our task. We will not dwell on this issue any further. The interested reader can find further information on it (and indeed on the whole field of 3D display) in [15] and its references.

8.6. Pointers to Further Reading

Now we have come to the end of the book. Restrictions on the time of the author (and also consideration of the restrictions on the time of the reader) did not allow us to cover everything known regarding the geometry of digital spaces. The interested reader will find much relevant material in the books [21, 30, 34, 35, 42, 45]. There are also two very relevant special journal issues, the October 1992 issue of *Topology and Its Application* and the June 1996 issue of the *Journal of Mathematical Imaging and Vision*.

8.7. Exercises

8.1. Label all the voxels in Figure 2.1.5 and, assuming that the 1-voxels are just the two drawn in detail, describe the performance of the 1-Simply Connected Algorithm when started with one of the surfels between a 1-voxel and a 0-voxel.

8.2. Define the notion of the symmetric closure ρ^s of a binary relation on a set M and using this notion prove the following. If C is a λ-component of the set of bels B in a binary picture, for some bel-adjacency λ, then (C, λ_C^s) is a digital space. (We use λ_C to denote the binary relation λ restricted to C.)

8.3. Discuss the consequences of the previous exercise from the point of view of our theory's ability to handle digital pictures on object boundaries.

8.4. Describe the performance of the General Bel-Tracking Algorithm for the binary picture $\left(Z^3, \omega_3, f\right)$ represented by Figure 1.9.1 with II taking the role of i and using a first-in-first-out queuing discipline.

8.5. Show that if (V, π, f) is a binary picture such that the set B of bels is finite, λ is a bel-adjacency and i is a bel, then T is empty at the time the General Bel-Tracking Algorithm terminates.

8.6. Prove that during the execution of Artzy's Algorithm, for every dirty face f the list contains $2 - $ (the number of dirty faces which are not in the queue and are adjacent to f) copies of that face.

8.7. Supply the details of the proof in Case (ii) for Theorem 8.4.1.

8.8. Provide the details of the inductive proof of the equivalence of the outputs of the Algorithm for Flat Flies and the General Bel-Tracking Algorithm with λ defined using the basic digraph of Figure 8.4.2.

8.9. State and prove the analog of Theorem 8.4.1 for $\omega_N \delta_N$-boundaries.

8.10. State precisely the version of the General Bel-Tracking Algorithm for efficiently tracking $\gamma_N \omega_N$-boundaries.

8.11. State precisely the version of Artzy's Algorithm for efficiently tracking $\alpha_3 \omega_3$-boundaries.

8.12. Show that, for any two distinct elements c and d of the fcc grid F, $(c, d) \in \overline{\beta_1}$ if, and only if, there is an element of $Z_3 - F$ which is ω_3-adjacent to both c and d.

8.13. Let (c, d) be an arbitrary bel of the finite picture (F, β_1, f), O be the β_1-component of the set of 1-spels that contains c, and Q be the $\overline{\beta_1}$-component of the set of 0-spels that contains d. Show how to specify the bel (c^0, d^0) in the derived binary picture (Z^3, ω_3, f') so that, when the Tracking Algorithm for Cubic and FCC Grids terminates, $S_F = \partial_{\beta_1}(O, Q)$.

8.14. Discuss the validity of the following claim. Let (F, β_1, f) be a finite picture over the fcc grid and let (c, d) be a bel that is "visible from the outside." Let O be the β_1-component of the set of 1-spels that contains c and let Q, respectively $\overline{Q}$, be the β_1-component, respectively $\overline{\beta_1}$-component, of the set of 0-spels that contains d. Then the set of bels that is "visible from the outside" in $\partial(O, Q)$ is exactly the same as the set of bels that is "visible from the outside" in $\partial(O, \overline{Q})$.

8.15. Show that the assumption that the set of bels is finite is not essential for the results proved in this chapter. In particular, prove the following variant of Theorem 8.4.1:

For $N \geq 2$, let $(M, \bigcup_{1 \leq i < j \leq N} \rho_{i,j})$ be a basic digraph for Z^N such that $\rho_{i,j} \neq \emptyset$ for any $1 \leq i < j \leq N$. For the binary relation λ on the set of all bels B in the binary picture (Z^N, ω_N, f) which is as defined above, for any bel $o = (c, d)$ such that λ-component of B which includes o is finite, the General Bel-Tracking Algorithm terminates in a finite number of steps and at that time $S = \partial(O, Q)$, where O is the δ_N-component of the set of 1-spels which contains c and Q is the ω_N-component of the set of 0-spels which contains d.

8.16. Show that the areas of the faces of the (polyhedral) Voronoi neighborhoods of δZ^3 and F_ϕ are the same if, and only if, $\phi = 2^{0.25}\delta$. Show, further, that in this case the volumes of the associated rhombic dodecahedra are $2^{1.75}\delta^3$.

8.17. Show that the Tracking Algorithm for Cubic and FCC Grids can be adapted to an arbitrary cubic grid δZ^3 and to an arbitrary fcc grid F_ϕ.

Appendix
List of Symbols

sgn	Sign function on real numbers	p. 63
$s_{\{i,j\}}$	N-dimensional sign function applied to i and j	p. 75
$S_T(c)$	Smallest neighborhood of c in topology T	p. 84
$\widetilde{S}$	Reverse-oriented version of the surface S	p. 72
u^n	Unit vector in direction n	p. 185
$u \cdot v$	Inner product of u and v	p. 1
$\|v\|$	Norm of v	p. 1
Z	Set of integers	p. 2
Z^N	Set of N-dimensional row vectors of integers	p. 2
α_N	An adjacency on Z^N	p. 8
β	An adjacency on Z^2	p. 50
β_s	An adjacency on Z^N	p. 75
β_ϕ	An adjacency on the fcc grid F_ϕ	p. 46
$\overline{\beta_1}$	An adjacency on the fcc grid F	p. 197
γ	An adjacency on Z^2	p. 51
γ_n	An adjacency on Z^N	p. 147
δ	An adjacency on Z^2	p. 51
δ_N	An adjacency on Z^N	p. 16
δZ^N	An N-dimensional cubic grid	p. 2
ε	Edge-adjacency on the hexagonal grid H	p. 53
ε_e	An adjacency on Z^N	p. 76
κ_N	Khalimsky adjacency on Z^N	p. 87
λ	Bel adjacency	p. 181
μ_ψ	Fuzzy connectedness associated with ψ	p. 103
ν	An adjacency on Z^2	p. 67
ν_i	An adjacency on Z	p. 117
ϕ	An adjacency on Z^4	p. 42
χ	An adjacency on Z^2	p. 59
ρ_T	Adjacency induced by topology T	p. 84

ρ^*	Transitive closure of ρ	p. 181
ω_N	An adjacency on Z^N	p. 8
$\in$	Is an element of	p. 2
$\notin$	Is not an element of	p. 2
$\emptyset$	Empty set	p. 11
$\Leftrightarrow$	If, and only if,	p. 7
$\subset$	Subset	p. 8
$\oplus$	Modulo 2 addition	p. 138
∂H	Bounding plane of a half-space H	p. 95
$\partial(O, Q)$	Boundary between O and Q	p. 11, 59
$\partial_\rho(O, Q)$	Boundary between O and Q using ρ	p. 127

References

[1] E.A. Abbott. *Flatland, A Romance of Many Dimensions*. Little, Brown and Co., Boston, 1899.

[2] R. Aharoni, G.T. Herman, and M. Loebl. Jordan graphs. *Graph. Models Image Proc.*, 58:345–359, 1996.

[3] E. Artzy, G. Frieder, and G.T. Herman. The theory, design, implementation and evaluation of a three-dimensional surface detection algorithm. *Comput. Graph. Image Proc.*, 15:1–24, 1981.

[4] F.K. Athappilly, R. Murali, J.J. Rux, Z. Cai, and R.M. Burnett. The refined crystal structure of hexon, the major coat protein of adenovirus type 2, at 2.9 Å resolution. *J. Molec. Biol.*, 242:430–455, 1994.

[5] L.-S. Chen, G.T. Herman, R.A. Reynolds, and J.K. Udupa. Surface shading in the cuberille environment. *IEEE Comput. Graph. Appl.*, 5(12):33–43, 1985. (Erratum appeared in 6(2):67–69, 1986.).

[6] D. Cohen-Or, A.E. Kaufman, and T.Y. Kong. On the soundness of surface voxelizations. In T.Y. Kong and A. Rosenfeld, editors, *Topological Algorithms in Digital Image Processing*, pages 181–204. Elsevier, Amsterdam, 1996.

[7] H.M.S. Coxeter. *Introduction to Geometry*. John Wiley & Sons, New York, 1961.

[8] P. Duchet, M. Las Vergnas, and H. Meyniel. Connected subsets of a graph and triangle basis of the cycle space. *Discrete Math.*, 62:145–154, 1986.

[9] M. Farber. Bridged graphs and geodesic convexity. *Discrete Math.*, 66:249–257, 1987.

[10] G. Frieder, G.T. Herman, C. Meyer, and J. Udupa. Large software problems for small computers: An example from medical imaging. *IEEE Software*, 2(5):37–47, 1985.

[11] D. Gordon and J.K. Udupa. Fast surface tracking in three-dimensional binary images. *Comput. Vision Graph. Image Proc.*, 45:196–241, 1989.

[12] F. Harary. *Graph Theory*. Addison–Wesley Publ. Co., Reading, MA, 1969.

[13] G.T. Herman. *Image Reconstruction from Projections: The Fundamentals of Computerized Tomography*. Academic Press, New York, 1980.

[14] G.T. Herman. Discrete multidimensional Jordan surfaces. *CVGIP: Graph. Models Image Proc.*, 54:507–515, 1992.

[15] G.T. Herman. 3D display: A survey from theory to applications. *Comput. Med. Imag. Graphics*, 17:231–242, 1993.

[16] G.T. Herman. Oriented surfaces in digital spaces. *CVGIP: Graph. Models Image Proc.*, 55:381–396, 1993.

[17] G.T. Herman and H.K. Liu. Dynamic boundary surface detection. *Comput. Graph. Image Proc.*, 7:130–138, 1978.

[18] G.T. Herman and D. Webster. A topological proof of a surface tracking algorithm. *Comput. Vision Graph. Image Proc.*, 23:162–177, 1983.

[19] S.W. Hughes, T.J. D'Arcy, D.J. Maxwell, J.E. Saunders, C.F. Ruff, W.S.C. Chiu, and R.J. Sheppard. Application of a new discreet form of Gauss' theorem for measuring volumes. *Phys. Med. Biol.*, 41:1809–1821, 1996.

[20] C. Kittel. *Introduction to Solid State Physics.* John Wiley & Sons, New York, 7th edition, 1996.

[21] T. Y. Kong and A. Rosenfeld, editors. *Topological Algorithms for Digital Image Processing.* Elsevier, Amsterdam, 1996.

[22] T.Y. Kong. Justification of a type of fast anisotropic boundary tracker for multidimensional binary images. In *Proc. SPIE*, volume 1832, pages 7–12, 1992.

[23] T.Y. Kong, R. Kopperman, and P.R. Meyer. A topological approach to digital topology. *Amer. Math. Monthly*, 98:901–917, 1991.

[24] T.Y. Kong, A.W. Roscoe, and A. Rosenfeld. Concepts of digital topology. *Topology Appl.*, 46:219–262, 1992.

[25] T.Y. Kong and A. Rosenfeld. Digital topology: Introduction and survey. *Comput. Vision Graph. Image Proc.*, 48:357–393, 1989.

[26] T.Y. Kong and J.K. Udupa. A justification of a fast surface tracking algorithm. *CVGIP: Graph. Models Image Proc.*, 54:162–170, 1992.

[27] V.A. Kovalevsky. Discrete topology and contour definition. *Pattern Recog. Lett.*, 2:281–288, 1984.

[28] L. Latecki. Topological connectedness and 8-connectedness in digital pictures. *CVGIP: Image Understanding*, 57:261–262, 1993.

[29] S. Matej and R.M. Lewitt. Efficient 3D grids for image reconstruction using spherically-symmetric volume elements. *IEEE Trans. Nucl. Sci.*, 42:1361–1370, 1995.

[30] S. Miguet, A. Montanvert, and S. Ubéda, editors. *Discrete Geometry for Computer Imagery.* Springer-Verlag, Berlin, 1996.

[31] J.R. Munkres. *Topology: A First Course.* Prentice–Hall, Englewood Cliffs, NJ, 1975.

[32] F. Natterer. *The Mathematics of Computerized Tomography.* John Wiley & Sons, Chichester, England, 1986.

[33] M.H.A. Newman. *Elements of the Topology of Plane Sets of Points.* Cambridge University Press, Cambridge, England, 2nd edition, 1951.

[34] Y. L. O, A. Toet, D Foster, H.J.A.M. Heijmans, and P. Meer, editors. *Shape in Picture: Mathematical Description of Shape in Grey-Level Images.* Springer-Verlag, Berlin, 1994.

[35] T. Pavlidis. *Algorithms for Graphics and Image Processing.* Computer Science Press, Rockville, MD, 1982.

[36] D.P. Petersen and D. Middleton. Sampling and reconstruction of wave-number-limited functions in *n*-dimensional Euclidean spaces. *Inform. and Control*, 5:279–323, 1962.

[37] K. Preston, Jr. Multidimensional logical transforms. *IEEE Trans. Pattern Anal. Mach. Intell.*, 5:539–554, 1983.

[38] A. Ralston and E.D. Reilly, editors. *Encyclopedia of Computer Science.* Van Nostrand Reinhold, New York, 3rd edition, 1993.

[39] A. Rosenfeld, T.Y. Kong, and A.Y. Wu. Digital surfaces. *CVGIP: Graph. Models Image Proc.*, 53:305–312, 1991.

[40] M. Senechal. *Quasicrystals and Geometry.* Cambridge University Press, Cambridge, England, 1995.

[41] J.K. Udupa. A unified theory of objects and their boundaries in multidimensional digital images. In *Proc. Comput. Assisted Radiol., CAR'87*, pages 779–784, Berlin, 1987.

[42] J.K. Udupa and G.T. Herman, editors. *3D Imaging in Medicine.* CRC Press, Boca Raton, FL, 1991.

[43] J.K. Udupa, H.-M. Hung, and K.-S. Chuang. Surface and volume rendering in 3D imaging: a comparison. *J. Digital Imag.*, 4:159–168, 1991.

[44] J.K. Udupa, L. Wei, S. Samarasekera, Y. Miki, M.A. van Buchem, and R.I. Grossman. Multiple sclerosis lesion quantification using fuzzy-connectedness principles. *IEEE Trans. Med. Imag.*, 16:598–609, 1997.

[45] K. Voss. *Discrete Images, Objects, and Functions in Z^n*. Springer-Verlag, Berlin, 1993.

Index